U0134989

50 大商業思想家論壇

Leadership

當代最具影響力 13位大師談領導

史都華‧克萊納 Stuart Crainer
德‧迪樂夫 Des Dearlove —— 著

譚天——譯

THINKERS
50

CONTENTS
目　錄

前言

領導
千年的議題

近年來，我們訪談了許多來自全球各地的人，請他們針對「領導」這個議題表示看法。這些人包含了跨國企業經理人、腦外科醫師、日本企業董事會、非營利組織領導人、足球隊教練、社會企業家、主廚、執行長、執行長候選人、企管碩士班學生、教育界領導人、教師等。

有意思的是，這些受訪者對這個議題幾乎都有看法。他們的見解之細、之精，屢屢讓我們稱奇不已。他們都曾讀過、深思過這個議題。領導的景觀已經改變。過去，領導只是少數軍事與政治首腦的專利；現在，領導已經民主化了，在我們的日常生活中，方方面面都與領導力有關。

隨著我們對領導的理解與實踐範疇不斷擴展，它所包含的能力範疇也愈來愈廣。現在的領導，用「命令」（command）與「控制」（control）幾個簡單、令人望而生畏的字，已經不再能一語道盡，它涉及了許多層面。領導涉及了感覺、情緒、追隨者，以及受領導人行動感召而有所行動的人。

不久前，我們在倫敦溫布萊體育館（Wembley Stadium）觀賞一場足球大賽。在中場休息時間、循例買漢堡果腹的時候，我們看到賣漢堡男子的衣服上，醒目地繡了「Team Leader」（團隊領導人）兩個英文字。中場休息時

間吵雜的人聲逐漸平息，我們找機會趨前問他：「我們看到你襯衫上那幾個字。依你看，什麼是領導？」他不假思索回答道：「四個字：以身作則。」

　　漢堡小販這話，印證了我們的一項認知：領導範疇無遠弗屆，即使在意想不到的地方也能見到領導。華倫・班尼斯（Warren Bennis）是領導思想家，曾替幾位美國總統工作，也是領導學大師。有一次我們與班尼斯會面，並問他如果他要問世界領導人一個問題，他會問什麼？班尼斯回答：「你如何學習？」

　　這本書收錄了我們與各界領導人和舉世聞名的領導思想家訪談的內容，包含了班尼斯、吉姆・柯林斯（Jim Collins）、席尼・芬克斯坦（Sydney Finkelstein）、史都華・費德曼（Stewart Friedman）、羅伯・高菲（Rob Goffee）、馬歇爾・葛史密斯（Marshall Goldsmith）、芭芭拉・凱勒曼（Barbara Kellerman）、拉克希・庫拉納（Rakesh Khurana），以及莉茲・梅隆（Liz Mellon）等人。我們的目的是讓讀者直接接觸、了解領導的基本面，以及有關領導的最新思考。身為領導人，一定有許多需要學習的東西。

史都華・克萊納（Stuart Crainer）與德・迪樂夫（Des Dearlove）
「50大商業思想家」（Thinkers50）創辦人

第 1 章

領導學的由來

每年，以領導與培養領導力為題而出版的新書，何止成千上萬。一疊又一疊的紙本與數不清的網頁，也義無反顧投入這個千百年來令領導人、未來領導人與追隨者同樣迷戀的主題。儘管有這麼多與領導有關的思考與討論，偉大領導的祕密，卻似乎始終遙不可及。如何才能成為卓越的領導人？每個人都有成為領導人的潛力，還是說它只是少數天縱英明之士的專利？為什麼民眾會在危急、艱困之秋追隨某個人，而不追隨另一個人？

這些都是世世代代以來，令人大惑不解的疑問。為了了解偉大的領導人之所以能讓人矢志效忠、無怨無悔的原因，我們從西元前 1 百年羅馬的凱撒大帝（Julius Caesar），一路研究、討論、分析到近年的史蒂夫·賈伯斯（Steve Jobs）。但我們仍在尋找答案，有效領導的神奇配方仍是不解之謎。

就連「什麼是領導？」這個最基本的問題，我們也不得其解。如果向 1 百位主管提出這個問題，你很可能會得到 1 百種答案。即使向窮畢生之功、鑽研這個議題的專家及學者請教，你仍然得不到一個大家都能接受的明確答案。管理大師湯姆·彼得斯（Tom Peters）曾說：「想在尋找領導新規則這件事上有任何進展，我們不如把話說在

前頭：關於領導這件事，沒有一種放諸四海皆準的入門模式。領導守則第一條：一切都得視情況而定。」

當然，許多權威專家對此提供了不同意見及看法，我們在這本書裡收錄了一些這類意見及看法。比方說，拿破崙（Napoleon）就曾經說過：「領導人是一種買賣希望的生意人。」想必他這話是親身體驗的有感而發。幾近兩個世紀以後，領導學大師班尼斯也指出：「經理人是把事情做對的人，而領導人是做對的事情的人。」

領導真的那麼重要嗎？顯然，它的確是。倘若這個世界沒有了領導，會變成什麼樣子？靜下心來想想，領導為這個世界帶來了什麼？它對你的日常工作、對你的家庭生活，有什麼影響？

以領導一個組織為例，領導一個大型組織，從來沒有像今天這麼不容易、這麼變化多端。就許多個案而言，報酬也從來沒有像今天這麼豐厚。此外，領導大型組織的工作，也從來沒有像今天這樣招徠這麼多批判。近年來，世界各地身敗名裂的執行長多不勝數，領導人的能力與道德標準，也因此又一次成為世人潛心探討的議題。就算領導還沒有出現全面敗壞的通盤性危機，但十幾年來的事件已

經足夠讓我們心驚膽戰，迫使我們重新思考領導理論、重新討論領導新做法。

在現代這個世界，通訊手段既迅速又四通八達，經濟與社會早已全球化、牽一髮而動全身，單單一位企業領導人的行動，可能影響好幾十億人的福祉。他們可能造成公司瓦解、市場崩盤，可能引發全球不景氣，導致一些國家的街頭暴亂。又或者，在千金難求的民望加持下，他們能為社會帶來成長與繁榮。

一些公司的執行長權勢之大，足以媲美國王、總統，有如將軍與總理。既然擁有如此大權，當然也得負起大得驚人的責任。而且，就算他們不是巨型跨國公司的執行長，企業無分大小，只要跨上全球商務，企業領導人都很重要，也都有影響力。他們或許沒有那種迫使國家向他們屈膝的權勢，但絕對能讓追隨他們的人過得更幸福、讓組織表現得更好、讓客戶更滿意；絕對能成就個人輝煌，也能為國家繁榮貢獻一己之力。

也因此，儘管有這麼多各說各話的理論與構想，領導人與領導術的研究仍然非常重要。想知道領導理論的發展，首先得了解領導的本質。理論與實踐原本就是盤根錯

節、難分難解的。

▶ 從古代到現代

　　那麼，領導理論的演進與它的研究情況如何？千年來，領導一直是個讓人迷戀的議題。早在古希臘時代，詩人荷馬（Homer）就寫過史詩，歌誦阿基里斯（Achilles）與奧德修斯（Odysseus）這類英雄的事蹟。希臘歷史學家布魯達克（Plutarch）也寫過《平行的生命》（*Parallel Lives*）一書，以傳記的方式記述凱撒等羅馬皇帝與其他名人的事蹟。在 19 世紀的維多利亞時代，湯瑪斯‧卡萊爾（Thomas Carlyle）也在他的《論英雄、英雄崇拜和歷史上的英雄業績》（*On Heroes, Hero-Worship, and the Heroic in History*）中，一一剖析了拿破崙等大人物的個性特質。

　　一般來說，早期領導理論主要集中在 3 個廣大的領導層面。第一個層面討論領導人的人格、特質與個性，也就是領導人的一般習性。第二個層面以領導人的行動，以及他們扮演的不同角色為主，談論領導人的作為與行事風格，而不是他們的個性與特色。第三個層面則以領導概念為重點，認為領導應視環境與背景不同而異，不同的情勢

需要不同的領導風格。

早年研究領導的人，一心一意關注的只是權勢與影響力。在他們眼中，領導是一種權力功能，是一種運用政治與影響技巧才能做到的東西。它曾讓尼可洛・馬基維利（Niccolò Machiavelli）熱衷不已──馬基維利是佛羅倫斯外交官，是 15 世紀文藝復興時代領導議題暢銷書《君王論》（*The Prince*）的作者，也是早期研究人性、特別是領導的學者。馬基維利的論點頗有可議之處，他強調欲達目的、不擇手段，極力主張有效領導必須權謀與威嚇並用。他說：「政治與道德無關。」直到今天，仍有許多人對他的觀點深信不疑。

到了幾世紀後的 1950 年代，社會心理學家約翰・法蘭奇（John French）與伯特倫・雷文（Bertram Raven）檢驗領導、權力與影響力之間的關係。領導以權力為基礎，但權力從何而來？他們為領導人找出 5 個權力來源：獎、懲（獎勵與懲罰他人的能力）；威信（一個人享有的聲望）；法權（一個人因為在組織內的地位而取得的權力）；專業能力（一個人根據專門知識而享有的權力）。

▶ 與生俱來或後天培養？

在領導這個議題上，問得最多的一個問題，大概就是：「領導人是天縱英明，還是可以後天培養？」領導人所以能夠領導，靠的是天賦資質，還是後天的訓練與努力？有很長一段時間，大家都認為領導人有與生俱來、沒辦法教、也學不會的資質，這就是 19 與 20 世紀初期流行的「偉人論」（great man theory）──但不包括「偉大女性論」（great woman theory），當年女性領導人的成功一般都遭到漠視，即便在之後許多年，這種現象依然屢見不鮮。

與「偉人論」觀點密切相關的「特質論」（trait theory），是領導世界的主流理論。這派學術主張找出普世性的領導特性或資質，認為偉大的領導人都有某些共同的人格資質、特色與個性，而這些資質、特色與個性就是領導人的標記。也因此，就理論而言，只要研究大量領導人，就能找出他們的共同資質。1980 年代中期，班尼斯就以這個理論為基礎，展開他對美國領導人的研究（詳見第二章）。班尼斯想為有效領導整理出一套法則，他以各別個性為準，針對美國男女領導人提出一種領導觀。

隨著時代變遷，「偉人論（與偉大女性論）」及「特

質論」逐漸失寵。但儘管這些理論曾遭致無數批判，研究領導的人仍往往忍不住見獵心喜，而將一堆領導人擺在一起進行調查，想找出他們的共同特性，將它們延伸運用到整體領導世界。

在 1960 年代與其後的十幾、二十年，領導學研究重心轉移到領導人的領導方式，也就是領導人的行動與行為。舉例來說，管理理論家羅伯・布雷克（Robert Blake）與珍・穆頓（Jane Mouton）研擬出一套管理座標，根據風格（特別是對事與對人的風格）將經理人分門別類。這套管理座標將管理風格從 1/1 一直排到 9/9，分為 5 大類：1/1 是什麼都不做的經理人，這類經理人幾乎沒有任何作為，無論對人或對事都漫不在乎；9/9 的領導人不僅擁有絕佳的組織技巧，還非常了解怎麼激勵部屬。

到了 1970 與 1980 年代，英國思想家約翰・艾戴爾（John Adair）將重心轉移到關鍵性領導功能（包括規劃、創意、管控、支援、訊息傳遞與評估）與領導責任區（任務編組、團隊與個人）。艾戴爾的「行動中心領導」（action centered leadership, ACL），是一種比較著重實用的領導評估概念，認為領導人的職責主要在於完成任務、建立團隊和激勵團隊成員。

除此之外，還有人對領導風格或功能，另外提出一些看法。例如，有人將領導人區分為指揮式（directive）與參與式（participative）領導人。指揮式領導人下命令與指示，代替他們的團隊做決定，並指望部屬遵令行事。相對而言，參與式領導人以一種較具磋商性的決策過程打動部屬，讓部屬更加投入。

▶ 環境背景才是王

領導學者也就情勢或環境對領導力扮演的角色進行探討。有些領導人在一段期間很有效，但在進入另一段期間之後就不行了。例如，溫斯頓・邱吉爾（Winston Churchill）在戰時是有效的領導人，但在承平時期他的領導就走樣了。或許情勢與環境背景不同，需要的領導風格也不同，而這就是「情境論」（situational theory）的要旨。有人從「情境論」衍生出「權變論」（contingency theory），考量情境變數，找出在某些環境與背景下最適當的領導風格。

如果談到情境領導，就不能不談保羅・赫西（Paul Hersey）與肯・布蘭佳（Ken Blanchard）兩人。赫西曾擔任

領導學教授，是《情境領導人》（*The Situational Leader*）一書的作者。而以《一分鐘經理人》（*The One Minute Manager*）一書享譽全球的布蘭佳，目前擔任肯‧布蘭佳公司（Ken Blanchard Companies）的精神長。兩人在 1969 年合寫《組織行為管理》（*The Management of Organizational Behavior*）時，首創情境領導模式。這套理論原本稱為「領導生命周期論」，在 1977 年換了一個比較沒那麼誇張的名字，改稱情境領導論。

布蘭佳與赫西找出 4 種能運用在不同情境的領導風格：命令（telling）、推銷（selling）、參與（participating）與授權（delegating）。「命令」是一種專制的領導風格，當部屬似乎不能或不願做他們該做的事情時，這種風格有效；「推銷」有時也可視為一種教練式領導；當領導人與追隨者共享決策過程，而領導人扮演一種促動角色時，「參與」風格很管用；領導人確立任務之後，把完成任務的責任交給部屬，就是「授權」。

另一方面，心理學家傅雷德‧費德勒（Fred E. Fiedler）提出一套權變領導模式。根據他的模式，領導效率與兩項因素有關，這兩項因素分別是領導風格（leadership style）與情境控制（situational control）──領導人因情勢所需而享有

控制力與影響力。這種控制力與影響力，又取決於另幾項因素，包括領導人與追隨者之間的關係、任務是不是結構性任務，以及領導人在組織內的權力有多大等。

▶ 變革型領導

或許在領導學界造成最大震撼的，是政治學家詹姆斯・麥格雷戈・伯恩斯（James MacGregor Burns）在 1970 年代末期推出的作品。伯恩斯提出一個概念，認為領導有兩種對比鮮明的風格，一是「處理型領導」（transactional leadership），一是「變革型領導」（transformational leadership）。在第一種風格中，領導人與追隨者之間有一種互惠關係，兩方的需求都一樣。相形之下，變革型領導講究的是雙方溝通，相互了解對方的動機，以建立一種彼此結合、相互激勵的新關係，這將領導人與追隨者之間的互動推上一種新層面。

自 1970 年代以來，變革型領導蔚為主流。無數學者投入變革型領導研究，在不同層面上提出各種觀點。首先要提的是伯納德・巴斯（Bernard Bass），他認為變革型領導有 4 項要件：理想化影響力（idealized influence）、

感召力（inspirational motivation）、知識啟發（intellectual stimulation）與個別化關懷（individualized consideration）。「理想化影響力」來自領導人的道德與倫理標準：領導人身為一種角色典範，為追隨者崇敬、仰慕；「感召力」刺激追隨者，讓他們追隨領導人、採取同一目標；「知識啟發」鼓勵獨立思考、爭議、討論、理性思考與解決問題；領導人親自關心追隨者、為追隨者提供建議，就是「個別化關懷」。

巴斯提出的這項概念，引起了廣泛的興趣。處理型領導人只重視「我為你這麼做，你就得這麼做來回報我」，只知道運用獎懲手段完成任務；感召型領導人卻講究個人情緒面，比起處理型領導人自然受歡迎得多。

▶ 逆轉勝的幕後推手

近年來，領導理論又開始重拾改變議題。舉例來說，麻省理工學院（MIT）的艾加‧席恩（Edgar Schein），與之後哈佛商學院（Harvard Business School）的約翰‧科特（John Kotter）與羅莎貝‧摩絲‧肯特（Rosabeth Moss Kanter），就都曾潛心鑽研過這個議題。

肯特觀察精於因應改革的領導人（即所謂的改革大師），對扭轉乾坤式的領導也有詳盡、透澈的研究。在對幾件反虧為盈、起死回生式的領導個案進行研究之後，她認為資訊與關係是重要因素。一位率領組織振衰起敝的領導人，必先促成一種態度與行為上的心理轉變，才能導致組織復甦。她在這種振衰起敝的過程中，找出 4 項基本要素：鼓勵對話、培養尊敬、促進合作與刺激創意。

與這種概念相關的是「引爆點式領導」（tipping point leadership），因麥爾坎·葛拉威爾（Malcolm Gladwell）所著的《引爆趨勢》（*The Tipping Point*）一書而爆紅。葛拉威爾發現，時裝的流行竟與傳染病的擴散及傳染病科學近似，並非一種有秩序的過程。葛拉威爾表示：「構想、產品、訊息與行為的傳播，就像病毒擴散一樣，只要少數幾個帶原者，就足以引發一場文化性的大感染。新構想的進展很快就會循一條迅速攀高的曲線不斷上升，直到抵達臨界質量的『引爆點』為止。」

有效率、有創意的領導，也能以同樣方式引爆。歐洲工商管理學院（INSEAD）教授金偉燦（W. Chan Kim）與蘭妮·莫伯尼（Renée Mauborgne）提出「引爆點式領導」構想，舉出前紐約市警局長威廉·布拉登（William Bratton，

見第八章）的事例，為這個想法帶來令人信服的佐證。

▶ 領導人的EQ

此外，心理學者、曾擔任《紐約時報》（*The New York Times*）記者的丹尼爾・高曼（Daniel Goleman），也以大衛・麥克里蘭（David McClelland）與霍華德・嘉納（Howard Gardner）的研究成果為基礎，提出領導人單憑智商（IQ）還不夠，還要有情緒智商（EQ）的構想。麥克里蘭是美國心理學理論學者，曾協助建立「職能模型」（competencies model），是高曼在哈佛大學的授業恩師。嘉納是發展心理學家，在哈佛大學教育研究所擔任認知發展心理學教授，是「多元智能理論」（theory of multiple intelligences）的創始人。高曼認為，經理人如果想成為有效的領導人，就需要了解並管理他們本身的情緒與人際關係。

高曼在《打造新領導人》（*Primal Leadership*）一書中，主張培養有 EQ 的領導人。他與共同作者理查・波雅齊斯（Richard E. Boyatzis）和安妮・瑪琪（Annie McKee）在書中指出，情緒智商有自我感知、自我管理、社會感知與關係

管理等 4 大領域，並說明這 4 種 EQ 如何形成不同風格的領導。根據這本書的說法，這 4 大領域就是情緒智商的全部內容，領導人只要能掌握這 4 大領域，就能充分發揮領導效應。

最後，到了 21 世紀，有些學者在領導議題上的做法更加激進。部分學者認為領導是跨團隊的課題，如賓州大學華頓商學院（The Wharton School）的凱薩琳・克蘭（Katherine J. Klein）便花了 10 個月的時間，在巴爾的摩休克創傷中心（Baltimore Shock Trauma Center）研究醫療團隊運作。她在近距離、親眼觀察領導作為之後，對領導產生一套獨特的看法。克蘭認為，領導「是一種系統或一種結構，它不是一種個人特性，而是組織或單位整體才有的特性」。

創傷醫療團隊的工作壓力奇大，環境混亂、吵雜、恐怖、緊張不說，決策略有錯失，時間稍微延誤，幾秒鐘之差就會導致生死殊途的後果。克蘭說，在這種工作條件下，領導是「一種角色。更明確地說，是充滿活力，因社會生活而有力、同時也受社會生活束縛的一套功能。在這個團隊中，長時間據有專業權威要職的許多個人，負責履行這套功能」。

在這類情境中，領導是組織或單位的「常規、慣例與角色詮釋」的產品。領導人的功能因環境不同，也與其他角色的功能不同。克蘭認為，領導有 4 種關鍵性功能：提供策略性方向、監控團隊表現、指導團隊成員、在必要時候伸手施援。根據克蘭的研究成果，組織應該建立結構，例如確立角色、明訂規範等，為任何一位擔任領導職的人提供必要支援，不必在遴選優秀領導人上花費太多工夫。

還有人認為，在組織性民主的企業中，員工是公民、不是單純的員工，而領導的目的就在於協助員工實現他們的抱負。這種理論以倫敦商學院（London Business School）林達・葛瑞騰（Lynda Gratton）教授為代表，葛瑞騰在她寫的《民主企業》（*The Democratic Enterprise*）一書中描述了這類場景。

▶ 百家爭鳴

儘管有這麼多人、這麼多理論、這麼多著作，投入領導這門課題，但一種大家都接受的明確領導定義，卻始終沒有出現。事實上，有關領導的理念，是愈來愈零碎了。傳統的領導概念，如資質與風格等仍然重要，英雄、名流

一般的執行長，在一些組織仍屬常態。同時，沉默領導、真誠領導（authentic leadership）、追隨理論，以及分布式領導（distributed leadership）等理論，也成為矚目焦點。

　　不過，無論怎麼說，我們對領導的了解，已經走出指揮式領導與威權領導的範疇，進入更加重視互動、重視自願接受領導的領域。過去，領導總是聚焦在執行長與高級主管、將領與總統、國王與王后；今天的領導，不再局限於組織高層，我們一般都相信，組織各階層、各角落都有領導人，而且這些領導人都像高層領導人一樣，需要同樣的支持與關切。

　　我們可以相當肯定地說，新的領導理念還是會繼續出現。世人雖早在千百年前已經開始探討領導，但領導學領域仍在持續演變中。我們要在後續幾章，針對百家爭鳴的現代領導理論抽絲剝繭，釐出一番頭緒。

第二章　領導的大考驗

　　班尼斯說，領導是切身、刻骨體驗的結果。雖說年輕未必是身為領導人的障礙，但年輕領導人如果能在經歷嚴苛考驗之後毫髮無傷，勝算就更大了。班尼斯告訴我們，所謂的「大考驗」（crucible）指的是「與過去完全不同的

事件或考驗，當事人必須通過這些事件或考驗，並且從中記取教訓，以期學習、成長與領導」。

但年輕領導人面對的挑戰是，這種「大考驗」並不多見，而且無法以人工複製。對許多年長的領導人而言，二次大戰與 1930 年代的大蕭條就是「大考驗」，他們的價值觀透過這些考驗而成形。下一代的領導人若想鑄造他們的領導才能與承諾，還得尋找其他經驗才成。

第三章　第五級領導

根據柯林斯的理論，領導人不僅需要了解自己與其他人的情緒，還必須保持謙卑。柯林斯著有 2001 年暢銷書《從 A 到 A+》（*Good to Great*），也是 1994 年暢銷書《基業長青》（*Build to Last*）的共同作者。他提出「第五級領導」（level 5 leadership），認為領導人需要兼具無私、謙卑與鋼鐵般意志的資質。這類領導人一般都是「沉默型領導人」，不是華爾街標榜的那種飛揚跋扈、魅力十足的英雄人物。對領導而言，謙卑是最後一個有待耕耘的領域。

第四章　真誠領導

如果說英雄式領導「出局」，真誠領導絕對稱得上已經「當道」。為解決明星領導人濫用領導造成的弊端，真

誠領導應運而生。我們不可能都成為英雄，但我們都能對自己真誠；領導是每個人都可以做到的事。

就某部分而言，真誠領導也反映了對英雄式領導與特質論的反撲。主張真誠領導概念的代表人物，包括曾擔任醫療產品公司美敦力（Medtronic）執行長的比爾・喬治（Bill George），與高菲、賈瑞斯・瓊斯（Gareth Jones）這類商學院學者。

最懂得善用真誠領導的領導人，能將他們已經具備的特質發揮到極致。他們知道怎麼靠自己的長處領導，也了解自己的短處。不過，這些特質必須真實、具備關鍵性、為他人所認可，才能派上用場。真誠領導談的，絕對不是套用其他成功領導人的風格或特質。

真誠領導需要反思與高度的自我認知，這些都是自我發展的必要階段。走捷徑省略這些階段的領導人，很可能扮演與自己價值觀或信念不符的錯誤角色。而扮演錯誤角色的領導人，很可能對組織造成重創，特別是如果他們運用這種錯誤角色與領導地位來掩飾自己的缺陷時，後果尤其嚴重。

第五章　魅力領導與黑暗面

　　這一章要探討的是英雄式與魅力領導的優點，以及它的黑暗面。1990 年代末期，麥克‧麥考比（Michael Maccoby）發現公司高層領導人的個性出現重大轉變。他發現，新一代商界領導人渴望聚光燈，並注意到「這些改變今日產業面貌的執行長們，有一種全新、大膽的風格」。根據麥考比的觀察，這些趾高氣揚、不可一世的領導人，與西格蒙德‧佛洛伊德（Sigmund Freud）所謂的自戀型人格類型非常近似。「自戀型領導」（narcissistic leadership）雖然未必是件壞事，但很容易成為壞事。

　　「魅力型領導」（charismatic leadership）與自戀型領導有關，提出這項概念的人，包括社會學家馬克斯‧韋伯（Max Weber），以及克雷蒙‧麥肯納學院（Claremont McKenna College）的傑伊‧康格（Jay Conger）教授等幾位領導理論學者。

　　魅力型領導人在歷史上占有重要地位，如拿破崙、邱吉爾、甘地都是這類領導人。過去，人們認為魅力是非常好的領導特質，甚至有人認為沒有魅力的人不能領導，但到了今天，魅力對領導的好處究竟有多少，已經愈來愈遭到眾人質疑。

第六章　追隨者

變革型領導理論的一個要點，是重視領導人與追隨者的關係。科特與約翰‧賈巴洛（John Gabarro）等兩位哈佛大學教授，在 1980 年發表一篇名為〈管理你的老闆〉（Managing Your Boss）的文章，將「經理人—老闆」的關係視為一種相互依賴的關係。他們指出，如果這種關係不是很好，身為追隨者的人應該花一些時間，耕耘更有效率的工作關係。

幾年後，顧問兼學者羅伯‧凱利（Robert Kelley）於 1988 年在《哈佛商業評論》（*The Harvard Business Review*）發表了一篇名為〈讚美追隨者〉（In Praise of Followers）的文章，將追隨者理念送上舞台中心。

哈佛大學約翰‧甘迺迪政府學院（John F. Kennedy School of Government）的凱勒曼教授，在《追隨力》（*Followership*）一書中問道，若是沒有好的追隨者，領導人又能有什麼成就？凱勒曼說，特別是在今天這個時代：「文化束縛削弱了有權、有勢、有影響力的人……追隨者的權力與影響力與日俱增，領導人的權力與影響力逐漸式微。」這個問題尤其值得我們深思。

第七章　領導人面對世界

領導並不鎖在一個企業與組織性真空裡，它能影響你的生活，與你整個一生重疊。但工作與生活應該如何平衡，這個問題對領導又有什麼意義？

第八章　領導人實踐

好的理論一定非常實用，領導是一門極度強調實用的學術，它會不會因為各式各樣的巧思，以及一些模糊不清的構想，而令人無所是從？

第 2 章

領導的大考驗

領導學之父

與談人｜**華倫・班尼斯**

　　領導人的一生總會有那麼一刻，在那刻要攀上高枝，或從管理階層躍居領導高位，或從團隊成員成為團隊領導人。對班尼斯來說，在二次大戰期間，在他還是歐洲戰場上一名年輕步兵軍官時，那一刻來了。班尼斯日後說，這次經驗是一次「大考驗」。

　　班尼斯如此寫道：「所謂的『大考驗』，就定義而言，是一種與過去全然不同的新體驗。經歷這種體驗的人，會因此產生一種全新、與過去不同的認同意識。」他日後在訪問領導人的時候，發現大考驗對塑造他們的領導特質影響很大。班尼斯表示：「我們在大考驗的過程中見到奇蹟出現。它就像一種魔法，將恐懼與苦難轉變為光榮與成功。這項過程即使不能創造領導力，也至少能夠開啟激勵、感召他人採取行動的能力。」

　　無論這些考驗是大戰期間的戰鬥、面對大災難的掙扎，或是人生旅途中其他是好或壞的重大事故，他們都能面對挑戰，思考因應之道、從中學習，並且因此成為更好的領導人。

▶ 親上火線

班尼斯住在加州,有一口潔白的牙齒,以及一身曬得古銅的皮膚。他看起來就像個領導人,20 年來,我們與他談過不知多少次話,他也總是非常熱誠地接受我們訪問。在討論「大考驗」的概念時,我們問他能不能自己創造一個大考驗。

可不可能為自己創造一個大考驗?

這是一個大問題。我認為大考驗一直都不斷地在出現。我們都經歷過大考驗,問題是我們在考驗結束後做了些什麼。我們從考驗中學到什麼?取得什麼智慧?我一直在想:我們的社會與教育機構為我們建立了多少能力,讓我們在組織或身為個人面對大考驗的時候,能夠了解情勢、認清狀況。問題不在於我們能不能創造大考驗;因為大考驗幾乎無時無刻都在出現。我們是不是將大考驗視為一個夢,早上醒來刷牙時,夢也煙消雲散,還是說我們會仔細思考這個夢,從夢裡學到東西?大考驗的情況也一樣:你得炒人魷魚、得被炒魷魚,公司把你派到一個你不喜歡的單位,或是你自認為遭到降級,而其實或許並非如此的崗位上。重要的是,一個組織能不能

從日常生活中不斷出現的大考驗中取得智慧，使它能夠有機性地透過經驗學習。我關心的是，如何利用日常生活中那些我們有時並不知覺的大考驗。

Is it possible for people to create their own crucible?

That's the big question. I think crucibles are created all the time. We all experience crucibles, but what do we do at the back end of them? Do we learn from them? Do we extract wisdom from them? I have been puzzling about how we create within our institutions the capacity to understand what goes on when organizations or individuals face crucibles. It isn't a question of how we create them; they happen almost all the time. Do we think of them as a dream, so that when we wake up and brush our teeth, they vaporize, or do we think about the dream and learn from it? It is the same thing with the crucibles: having to fire people, being fired, being shipped to an office you don't like, or thinking that you have been demoted when maybe you haven't been. It's a matter of how an organization can use the crucibles of everyday life and extract wisdom from them to make the organization learn organically from the experiences it's undergoing. My concern is, how do we use everyday crucibles that we're sometimes not conscious of.

———

這麼說，領導人必須設法尋求不確定性了？

你不能依法炮製一個像曼德拉（Nelson Mandela）被關在
羅賓島（Robben Island）的經驗，也不能複製約翰・麥坎
（John McCain）在越南成為戰俘被關的經驗，這些都是極
端的例子。

So leaders have to seek out uncertainty?

You can't create Mandela's Robben Island or John McCain's
experiences in Vietnam. They are extreme.

———

你活在什麼時代，就是什麼時代。總不能怪生不逢時
吧。

柯林頓總統（Bill Clinton）總難掩那一絲對其他幾任總統
的豔羨，因為他在職期間，沒有碰上可以讓他證明自己
的戰爭。老羅斯福總統（Teddy Roosevelt）的情況也一樣，
不過他處理過幾場小型戰事。

有人認為，今天這些領導人，都是從 1960 年代塑造出來
的。1960 年代對美國而言，是一個紛紛擾擾的年代，但他
們並沒有經歷過類似二次大戰或大蕭條這樣的大考驗。

看著這一代「奇葩」人物（geeks），你可以認為他們在911 事件過後應該定型了。這一代人的塑造早自 1989 年柏林圍牆倒塌、冷戰結束時已經開始，之後更受到全球資訊網（World Wide Web）普及的影響。所以，這不是一代人一代人的問題；它的影響期比較短。

You can't be held responsible for the era in which you live.

President Clinton was always slightly envious of other presidents because he didn't have a war to deal with that would let him prove himself. Teddy Roosevelt was the same, although he had a few minor skirmishes.

There is a view that the leaders who are around now were shaped by the 1960s, which was a fractious time in the United States. They didn't experience a crucible like World War II or the Depression.

You could look at this generation of geeks and say that their formative period ended at 9/11, but it started in 1989 when the Berlin Wall fell and the Cold War ended, and then there was the introduction of the World Wide Web. So it's not a generational thing ; it affects a shorter period.

———

如果說我們是環境的動物，這話是不是指對於環境這個問題，我們無能為力？

沒錯。約翰·賈納（John Gardner）是我們領導學界的一位老前輩，他在二次大戰期間曾在海軍陸戰隊服役，後來替詹森總統工作。他個性害羞、內向，但面對領導議題卻總是一馬當先。有一次，我在訪問他的同一週，也訪問了兩個年輕人，這兩個年輕人曾被迫解雇了 25 名友人。我問賈納，打仗與炒好友魷魚，哪件事更不安、惱人？賈納說，他不知道。

但遇上重大事件爆發，令你思索這世界會發生什麼變化時，這些事件與公司派你到另一個國家工作的事，自然大不相同。

If we are creatures of circumstance, doesn't that mean we are powerless?

That's right. One of our grand old men of leadership, John Gardner, had been a marine in World War II and worked with President Johnson. He was shy and introverted, but he was plunged into leadership. I interviewed him during a week when I also interviewed a couple of young guys who had had to lay off 25 of their friends from their business. I asked Gardner what he thought would create more angst, more emotional charge: being in the war or firing some of your closest friends. He was unsure.

But events on a cosmic scale, when you're thinking about what might happen to the world, are profoundly different from being sent to work in another country for your employer.

———

你在大戰期間的經驗，對你而言當然是一次大考驗。不過，你是不是因為那次經驗而自視為領導人？

我現在確實認為自己是一位思想領導人。我出身於一個非常貧窮的家庭，所以大戰結束後，我想到自己挺過了這場大戰，覺得自己身為青年軍官的表現也不錯。戰後我駐在法蘭克福，有一輛吉普車，還有一間公寓，生活相當寫意。我在這場大戰中，學到的是紀律與一種自控意識。我覺得我已經學會自我照顧，還想多學一些。當時我想，現在我可以面對人生了，但那時的我並不覺得自己是個領導人。一直到 1947 年 4 月，我才離開軍中。這段期間給我的影響很大，給了我許多原本不可能碰上的經驗。我原本非常害羞，也覺得自己是個很無趣的人，但在軍中這段經歷，讓我更加關心自己。我覺得自己成熟了。

Your experiences in the war were obviously a crucible for you, but did you emerge from that thinking of yourself as a leader?

I do think of myself now as a thought leader. I came from a very poor family, so after the war, I was thinking that I'd got through it and felt okay about what I had done as a young officer, and then I was stationed in Frankfurt after the war, and I had a Jeep and an apartment, and it was quite a good life. What I had learned was discipline and a sense of self-mastery. I felt that I took care of myself and was motivated to learn more. I thought, now I am ready to face life, but I didn't feel as though I was a leader. I stayed on until April 1947. It shaped me so much and pulled from me things I might not otherwise have experienced. I was very shy, and I felt that I was a boring human being, and then in the course of being in the army, I felt that I was more interesting to myself. It was a coming of age.

▶ 大戰過後

　　大戰結束後，班尼斯在安提亞克學院（Antioch College）念大學，師事道格拉斯・麥格雷戈（Douglas McGregor），麥格雷戈是 X 理論與 Y 理論動機概念的研創人。班尼斯後來隨麥格雷戈前往麻省理工學院，在麻省理

工學院拿到經濟學與社會學博士學位。之後，他在麻省理工學院任教，擔任組織研究系系主任。班尼斯說：「我早期的工作是『小團隊動力』（small group dynamics），這主要是社會心理學一塊經典領域。後來進入『T 群敏感度訓練』（T-groups, sensitivity training），之後再轉入社會系統。」

在 1950 年代投入「團隊動力」研究的班尼斯，在 1960 年代成為未來學者。他的著作，特別是 1968 年出版的《臨時社會》（*The Temporary Society*），探討了新組織的形式。班尼斯認為，組織需要建立「彈性形式」（adhocracies），掙脫階級與全無意義的紙上作業的桎梏與枷鎖。所謂「彈性形式」差不多是「官僚體制」（bureaucracies）的反義詞，後來艾文・托佛勒（Alvin Toffler）等人都用了這個名詞。

此時的班尼斯，既是紐約州立大學水牛城分校（State University of New York at Buffalo）副校長，又是辛辛那提大學（University of Cincinnati）校長。在為商界預測未來世界的同時，身為領導人的班尼斯，難免面對一些有時令人沮喪的現實。領導理論雖是他的熱情所繫，實踐成果卻並不能如他所期。

班尼斯表示：「在辛辛那提大學的時候，我發現自己之所以當這間大學的校長，為的是想透過地位取得權力。我想當一位大學校長，但我不想做校長的工作。我要的是影響力。最後，我這個校長當得不怎麼好。我望著窗外，心想在那裡割草的那個人，似乎比我更能控制他手上的工作。」

後來，班尼斯重拾他最喜歡的工作：教書、研究、顧問、寫作，以及發表有關領導的演說。今天，他是洛杉磯南加州大學馬歇爾商學院（Marshall School of Business）企業管理學特聘教授、管理與組織學教授，也是南加大領導研究所（Leadership Institute）創辦人。他同時也是哈佛大學甘迺迪政府學院公共領導中心（Center for Public Leadership）顧問委員會主席。

班尼斯說：「從我開始思考時，就在想著領導。這或許也是我作繭自縛。我的第一篇論文在 1959 年發表，談的就是領導。自 1985 年起，我的大多數作品都與領導有關。你營造一種品牌資產，自然而然，那個品牌資產與市場之間就會產生一種連結。有人談到領導，就會說那是班尼斯的東西。這使人生稍稍單純了些。」

▶ 最重要的本領是學習

1985 年，班尼斯與南加大未來研究中心（Center of Futures Research）創辦人兼主任伯特・納努斯（Burt Nanus），對全美最有名的 90 位領導人的人生進行研究、檢驗之後，合著了《領導人：主事的策略》（*Leaders: The Strategies for Taking Charge*）這本書。這本書蒐羅的名人來自各行各業，有麥當勞（McDonald's）創辦人雷・克洛克（Ray Kroc），以及來自商界、運動與藝術界的其他許多名流，甚至還有太空人尼爾・阿姆斯壯（Neil Armstrong）。

班尼斯說：「他們有的善用右腦思考、有的善用左腦思考；有的高、有的矮、有的胖、有的瘦；有的能言善道、有的不善言語；有的堅決果斷、有的怯懦優柔；有的衣著得體、有的不善穿著；有的重參與共享、也有的專制獨裁。」儘管有這麼多差異，但他們至少有一件事是一樣的，那就是都「善於掌控當前的混亂」。

班尼斯與納努斯從這些領導人中，歸納出 4 種共同能力：注意力、意義、信任與自我的管理。意義的管理涉及有效運用溝通技巧與科技，強化視覺效果，讓視覺栩栩如生。有效溝通涉及的技巧包括比擬、暗喻、生動的圖示、

情緒、信任、樂觀與希望。信任是「將追隨者與領導人結合在一起的情緒黏膠」。信任的建立有幾項要素，信任的基礎則是前後一致。

根據班尼斯的這項研究，領導人最後還有一種共同能力，就是「自我調度」（deployment of self）的能力。想成為一位好的領導人，需要下苦功，重點是要堅持不懈、要有自我認知，要能夠冒險、投入與挑戰，但最重要的是要學習。班尼斯說：「不斷學習的人，不會迴避失敗或錯誤。領導人最嚴重的問題，基本上就是少年得志，沒有機會從逆境與難題中學習。」

此外，領導人還具備「情緒智商」，也就是能正面看待自己。這種智慧的特徵是，能夠不嫌棄他人；能夠不翻舊帳、針對目前情況處理問題；能夠禮貌地對待每一個人；能夠信任他人，即使這麼做可能有危險也不在乎；能夠在沒有人不斷給予贊同與認可的情況下，依然勇於向前。

▶ 每個人對世界都能有所貢獻

班尼斯一再強調，領導力是可以學習的。在他的著作，特別是《領導人》這本書，都強調了這點：「每個人都必須在人生旅途中有一份真正的貢獻，去上班工作就是完成這項任務的一種重要工具。我愈來愈相信，每位領導人都可以建立一種人性社群，這社群經過不斷地演變，最後成為最好的組織。」

不過，為了達到這項目標，我們需要克服 5 個有關領導的迷思。首先，領導不是一種稀奇罕有的技巧。其次，領導是後天養成的，不是與生俱來的。第三，領導人大體上都是平凡或看起來平凡的人，而不是魅力十足之士。第四，領導不是組織高層的專利，無論置身任何階層，都可以領導。最後，領導的要旨不在於控制、指揮與操縱；它的要旨是以一種誘人的目標讓人全力以赴。

其後，班尼斯在《領導，不需要頭銜》（*On Becoming a Leader*）一書中，將注意力轉移到組織裡一個經常有人提出的問題：經理人與領導人有什麼不同？班尼斯說，這兩者有一種重大差異：「想在 21 世紀存活，我們需要的是新一代領導人，不是新一代經理人。其最大不同在於：領導人

能夠征服環境。我們周遭那一切變幻莫測、動盪不安與模糊不清的事物，讓我們備受威脅，稍有不慎就能讓我們窒息，而經理人碰上這些威脅只會投降。」

班尼斯提出一紙清單，說明經理人與領導人的基本差異：

- 經理人執行處理；領導人開創布新。
- 經理人是拷貝；領導人是原件。
- 經理人維持；領導人發展。
- 經理人注重系統與結構；領導人注重的是人。
- 經理人靠控制；領導人懂得激發信任。
- 經理人眼光短；領導人看得遠。
- 經理人會問「如何」與「何時」；領導人會問「什麼」與「為什麼」。
- 經理人只看近利；領導人會放眼未來，遠眺地平線。
- 經理人接受現狀；領導人挑戰現狀。
- 經理人聽命行事，是傳統好戰士；領導人有自己的主張。
- 經理人能把事情做對；領導人做對的事。

▶ 如何打造天才團隊？

到了 1990 年代中期，班尼斯開始探討團隊工作與合作領導的問題。在他 1997 年出版的《7 個天才團隊的故事》（*Organizing Genius*）中，班尼斯的注意力已經轉移到團隊力量、如何建立團隊，以及領導在團隊的角色等問題。他在書中描述了一些傑出團隊的成就，包括航太公司洛克希德（Lockheed）的祕密研究單位臭鼬工廠（Skunk Works）、發明原子彈的「曼哈頓計劃」（Manhattan Project）研究團隊，以及全錄（Xerox）那個舉世聞名的帕羅奧圖研究中心（Palo Alto Research Center）。

在一個好的團隊裡，領導人是不可或缺的重要因素。不過，一個團隊之所以傑出，主要靠的不是它的領導人，而是團隊全體成員以及他們的互動。再一次，班尼斯對英雄式領導人的概念提出質疑。他說，英雄式領導人不屈不撓、英勇非凡的概念，現在不僅已經過時，而且也不恰當。新一代領導人「是實事求是的夢想家，有屬於自己的遠見，但這些遠見並非遙不可及。頗具反諷意味的是，這類領導人的夢，只有在讓其他人放手一搏、表現特別優異的情況下，才能成真」。

　　領導人不應該用自己的一套標準強行領導，而應該找出一條適合團隊的領導之道。班尼斯指出：「無論如何，領導人總得找出一種適合團隊的領導風格。所謂的標準模式，特別是指揮、命令那套領導模式，已經不管用了。團隊負責人必須果決，但絕不能獨斷獨行。他們應該在不損及其他共事人自決的情況下做決定。如何設計、維持一種氛圍，讓其他人都能表白發揮，是領導人的創意之舉。」

　　差不多在出版《7 個天才團隊的故事》的同時，班尼斯對執行長以下的團隊領導人展開研究。在組織裡面，負責實際工作的一般都是這類領導人。班尼斯與喬治城大學（Georgetown University）客座教授大衛・漢南（David A. Heenan）聯手，對時任微軟執行長比爾・蓋茲（Bill Gates）的副手史蒂夫・鮑爾默（Steve Ballmer）進行觀察。兩人還一起研究了其他許多共同領導人，了解領導人與被領導人之間的關係，以了解組織概況。針對這個議題，兩人合著了《最佳拍檔》（*Co-Leaders: The Power of Great Partnerships*）一書，該書在 1999 年出版。

▶ 新一代領導人要有的 4 種能力

到了 2000 年代，班尼斯提出所謂「新領導」（new leadership）的主張。他說，現代價值鏈過於零碎，需要一種非常注重影響力的新類型領導。我們與班尼斯討論了「新領導」，以及他對團隊領導的心得。

你自認是個浪漫派嗎？

如果所謂浪漫派，指的是一個相信事在人為、一個樂觀的人，或許我真的是浪漫派。我認為，每個人都必須在人生旅途中有一份真正的貢獻，去上班工作就是完成這項任務的一種重要工具。我愈來愈相信，每位領導人都可以建立一種人性社群，這社群經過不斷地演變，最後成為最好的組織。

Do you see yourself as a romantic?

If a romantic is someone who believes in possibilities and who is optimistic, then that is probably an accurate description. I think that every person has to make a genuine contribution in life, and the institution of work is one of the main vehicles for achieving this. I'm more and more convinced that individual leaders can create a human community that will, in the long run, lead to the best organizations.

——

偉大的團隊需要偉大的領導人嗎？

想成就偉業，必須先有偉人。沒有偉大的領導人，偉大
的團隊也無法存在。但一種根深柢固的錯誤觀念也因此
出現，認為成功的組織是偉人身影的延伸。人生恐怕從
來不曾如此單純，單純到單憑一個人就能解決大多數重
大問題。沒有一個人的聰明才智，能夠勝過集體智慧。

Do great groups require great leaders?

Greatness starts with superb people. Great groups don't exist
without great leaders, but they give the lie to the persistent
notion that successful institutions are the lengthened shadow of
a great woman or man. It's not clear that life was ever so simple
that individuals, acting alone, could solve most significant
problems. None of us is as smart as all of us.

——

**所以，像影星約翰・韋恩（John Wayne）式的英雄已經是
歷史了？**

沒錯。獨行俠式的英雄已經做古。單打獨鬥、解決問題

的時代已經結束，新的創意、成就模式已經出現。像史蒂夫・賈伯斯與華特・迪士尼（Walt Disney）這類領導人會領導組織，在組織中營造自己的偉業。這種新領導人是實事求是的夢想家，他們有屬於自己的遠見，但這些遠見並非遙不可及。頗具反諷意味的是，這類領導人的夢，只有在讓其他人放手一搏、表現特別優異的情況下，才能成真。這類領導人一般很懂得招徠信徒，他們善於編織美夢，能將願景描繪得栩栩如生，讓人心悅誠服、跟著他們打拚。

So, the John Wayne type of hero is a thing of the past?

Yes, the Lone Ranger is dead. Instead of the individual problem solver, we have a new model for creative achievement. People like Steve Jobs and Walt Disney headed groups and found their own greatness in them. The new leader is a pragmatic dreamer, a person with an original but attainable vision. Ironically, the leader is able to realize his or her dream only if the others are free to do exceptional work. Typically, the leader is the one who recruits the others, by making the vision so palpable and seductive that they see it, too, and eagerly sign up.

———

但，這是不是有點不切實際？

說得沒錯。大多數組織都很麻木,職場生活也十分繁瑣
無趣。掙不開、走不脫。因此,這類新團體讓人精神
為之一振。偉大的團體不單只是結合一群第一流人才而
已,它還是一種奇蹟。我樂觀認為,只要放眼未來,相
信事有可為,我們都能有所長進。

But isn't this somewhat unrealistic?

True. Most organizations are dull, and working life is mundane.
There is no getting away from that. So, these groups could be an
inspiration. A great group is more than a collection of first-rate
minds. It's a miracle. I have unwarranted optimism. By looking
at the possibilities, we can all improve.

————

未來的領導人,需要具備哪些東西才能有效領導?

後官僚時代的組織,需要重新調整領導人與被領導人之
間的關係。今天的組織已經出現邦聯、網絡、群體、跨
功能團隊、臨時系統、特定任務編組、格子、模子、矩
陣等種種型態,五花八門,什麼都有,就是沒有那種由
上而下領導的金字塔式型態。新領導人會鼓勵部屬提出
健康的異議,會重視那些敢於說不的追隨者。

這樣的發展不表示領導已經步入尾聲。事實上，它表示
領導人如果想有效領導，就需要一種遠比過去更精緻、
更間接的新影響形式。智慧資產（腦力、知識與人的想
像力）已經取代資金，成為決勝的關鍵要素，這是新的
現實。面對這種新現實，領導人必須學習一整套新技
巧。我們今天還不了解這些技巧，我們的商學院也還沒
有教，有關的運作實例當然更是難得一見。新領導能不
能成功，取決於 4 種能力。

What will it take for future leaders to be effective?

The postbureaucratic organization requires a new kind of
alliance between the leaders and the led. Today's organizations
are evolving into federations, networks, clusters, cross-functional
teams, temporary systems, ad hoc task forces, lattices, modules,
matrices—almost anything but pyramids with their obsolete
top-down leadership. The new leader will encourage healthy
dissent and values those followers who are courageous enough to
say no.

This does not mark the end of leadership—rather, it indicates
the need for a new, far more subtle and indirect form of influence
if leaders are to be effective. The new reality is that intellectual
capital (brainpower, know-how, and human imagination) has

supplanted capital as the critical success factor, and leaders will have to learn an entirely new set of skills that are not understood, are not taught in our business schools, and, for all of those reasons, are rarely practiced. Four competencies will determine the success of new leadership.

———

第一種是什麼？

新領導人了解賞識人的重要性，也在這方面力求實踐。這類型領導人是人才的鑑賞家，他們比較不像人才創造人，更像是人才監護人。在新組織裡，他們一般不是最優秀、最有才氣的人。新領導人知人善用，就像有一具 Rolodex 旋轉式名片架一樣，懂得選拔人才，而且不怕雇用比他們強的人。我研究過許多表現傑出的團體，發現在大多數案例中，領導人都不是團體中最聰明或最敏銳的一員。迪士尼動畫片廠（Feature Animation）是迪士尼公司非常成功的部門，彼得・史奈德（Peter Schneider）是片廠負責人，領導了 1,200 位動畫師，卻不會畫畫。帕羅奧圖研究中心是第一部商業性個人電腦的發明地，曾經擔任中心負責人的鮑伯・泰勒（Bob Taylor）卻不是電腦科學家。美國企業家麥克斯・帝普雷（Max De Pree）曾說，好的領導人會「為了求取人才拋棄自我」，這話說得對極了。

What's the first?

The new leader understands and practices the power of appreciation. These leaders are connoisseurs of talent, more curators than creators. The leader is rarely the best or the brightest in the new organizations. New leaders have a smell for talent, an imaginative Rolodex, and are unafraid of hiring people who are better than they are. In my research into great groups, I found that in most cases, the leader was rarely the cleverest or the sharpest. Peter Schneider, president of Disney's colossally successful Feature Animation studio, leads a group of 1,200 animators. He can't draw to save his life. Bob Taylor, former head of the Palo Alto Research Center, where the first commercial PC was invented, wasn't a computer scientist. Max De Pree put it best when he said that good leaders "abandon their ego to the talents of others."

――

那第二種是什麼？

新領導人會不斷提醒追隨者，要他們注意重要的事。組織漸趨一同，想像力愈來愈僵化，員工也逐漸淡忘重要的事。「提醒你們的部屬，要他們注意重要的事。」這話說來簡單，但我每每在向領導人提供建議的時候，都

會提出這句話。一個強有力的願景，可以讓原本暮氣沉沉的工作，化為集體投入的壯舉。以曼哈頓計劃為證，美國陸軍從全國各地找來許多優秀的工程師，賦予他們特定任務。這些工程師就用當時（1943-1945 年）還很原始的電腦，做著能量計算與其他一些枯燥的工作。

但陸軍當局守口如瓶，怎麼也不肯告訴他們任何有關這項計劃的細節。工程師們不但不知道自己是在造一個可以讓大戰結束的武器，甚至不知道自己計算的是些什麼。當局只是告訴他們、要他們怎麼做，如此而已。工程師們照做了，做得很慢，也做得不怎麼好。當時負責督導這些技術人員的理查德‧費曼（Richard Feynman）說服他的上司，要上司告訴這些工程師他們在做些什麼，以及為什麼要做。當局終於批准揭密，曼哈頓計劃主持人羅伯特‧歐本海默（Robert Oppenheimer）於是發表專題演說，向他們說明計劃性質以及他們對計劃的貢獻。

「整個情勢完全變了，」費曼回憶道。「他們開始創造把事情做得更好的新辦法。他們改進了工作計劃，在晚上加班趕工。他們在晚上工作，不需要監工，什麼都不需要。他們了解一切，還研發了幾個案子供我們使用。」費曼估計，在技術人員知道曼哈頓計劃的重要性以後，工作進度「快了將近 10 倍」。

查爾斯・韓第（Charles Handy）在《適當的自私》（*The Hungry Spirit*）一書中說得好。我們都是「餓精靈」，渴望我們的工作有宗旨、有意義，渴望有超越自我的貢獻，領導人絕不能忘了提醒眾人，要他們注意重要的事。

Then, what's next?

The new leader keeps reminding people of what's important. Organizations drift into entropy and the bureaucratization of imagination when they forget what's important. It's simple to say, but that one sentence is one of the few pieces of advice I suggest to leaders: remind your people of what's important. A powerful enough vision can transform what would otherwise be routine and drudgery into collectively focused energy. Witness the Manhattan Project. The U.S. Army had recruited talented engineers from all over the United States for special duty on the project. They were assigned to work on the primitive computers of the period (1943-1945), doing energy calculations and other tedious jobs.

But the army was obsessed with security and refused to tell them anything specific about the project. They didn't know that they were building a weapon that could end the war or even what their calculations meant. They were simply expected to do the work, which they did slowly and not very well. Richard

Feynman, who supervised the technicians, prevailed on his superiors to tell the recruits what they were doing and why. Permission to lift the veil of secrecy was granted, and Robert Oppenheimer gave them a special lecture on the nature of the project and their own contribution.

"Complete transformation," Feynman recalled. "They began to invent ways of doing it better. They improved the scheme. They worked at night. They didn't need supervising in the night; they didn't need anything. They understood everything; they invented several of the programs we used." Feynman calculated that the work was done "nearly 10 times as fast" after it had meaning.

Charles Handy has it right in his book The Hungry Spirit. We are all hungry spirits craving purpose and meaning at work, to contribute something beyond ourselves, and leaders must never forget to remind people of what's important.

新領導人還需要具備什麼？

新領導人能為組織營造一種信任，並且維護這種信任。我們都知道新社會工作合約的條件已經轉變，沒有人還敢對任何組織付出終身忠誠或承諾。自 1985 年以來，約 25% 的美國勞動人口至少曾被解雇一次。組織與知識工

作者之間的關係，已因新社會合約而愈來愈曖昧不明，在這樣的時代，信任成為將員工與組織結合在一起的情緒黏膠。

「信任」這個字詞看起來雖不起眼，卻有很強的內涵，而且是一種極端複雜的東西。它的成分包括能力、堅持、關懷、公正、坦率與真誠，其中尤以真誠最為重要。抱負、能力與正直，是我們大多數人職場生活成敗的 3 大支柱，新領導人懂得取得這三者之間的平衡，營造信任。

What else does a new leader strive for?

The new leader generates and sustains trust. We're all aware that the terms of the new social contract of work have changed. No one can depend on lifelong loyalty or commitment to any organization. Since 1985, about 25 percent of the American workforce has been laid off at least once. At a time when the new social contract makes the ties between organizations and their knowledge workers tenuous, trust becomes the emotional glue that can bond people to an organization.

Trust is a small word with powerful connotations and is a hugely complex factor. The ingredients are a combination of competencies, constancy, caring, fairness, candor, and authenticity—most of all, the last. And the new leaders achieve that when they can successfully balance the tripod of forces that are working on and in most of us: ambition, competence, and integrity.

———

最後一種能力是？

新領導人與被領導人是親密的盟友。史蒂芬・史匹伯（Steven Spielberg）導演的電影《辛德勒的名單》（*Schindler's List*）之所以看了讓人震撼不已，正因為辛德勒的轉變。他原本是個窮困潦倒、低下卑劣的小混混，為榨取廉價猶太人勞工而移居波蘭，製造彈藥、以低成本賣給德國牟利。但在與他手下那些猶太人員工互動，特別是在與他的會計師雷文（Levin）交往之後，他開始轉變。此外，戰爭與大屠殺的邪惡，屢屢為他帶來的錐心刺骨之痛，也是促使他轉變的要因。片子演到倒數第二幕時，戰爭已經結束，納粹也已撤離了工廠，但在美軍抵達以前，囚犯贈送他一枚他們用製彈藥的貴金屬為他做的戒指。辛德勒一邊試著戴上這枚戒指，一邊忍不住落淚：「為什麼，為什麼你們要這麼做？如果不做這枚戒指，我們原可用這些金屬多救 3 個、4 個，或者 5 個猶太人。」他淚流滿面，駕車離去。

面對這樣的劇情，想以公正的態度進行分析很難。不過，儘管這是極端特殊、不凡的事件，它能將新領導人的精義描繪得淋漓盡致：只有偉大的團隊，只有能建立尊敬與尊嚴的組織，才能造就偉大的領導人。這類新領導人不會大聲叫嚷，但他們有一雙非常懂得聆聽的耳朵。這類後官僚組織不會有金字塔式、由上而下的階層結構。它們由活力與構想組建而成，領導它們的，是熱愛手邊工作的領導人，這些領導人會擁抱他們的同事，對於自己能不能揚名立萬並不在意。

And the last competency?

The new leader and the led are intimate allies. The power of Steven Spielberg's *Schindler's List* lies in the transformation of Schindler from a sleazy, down-at-the-heels, small-time con man who moves to Poland in order to harness cheap Jewish labor to make munitions that he can then sell to the Germans at low cost. His transformation comes over a period of time during which he interacts with his Jewish workers, most of all the accountant, Levin, but there are also frequent and achingly painful moments where he confronts the evil of the war and of the Holocaust. In the penultimate scene, when the war is over and the Nazis have evacuated the factory, but before the American troops arrive, the prisoners give him a ring that they have made for him from the precious metals that the workers used. As he tries to put the ring on, he begins crying, "Why, why are you doing this? With this metal, we could have saved three, maybe four, maybe five more Jews." And he drives off in tears.

It is hard to be objective about this scene, but, although this was a unique, singular event, it portrays what the new leadership is all about: that great leaders are made by great groups and by organizations that create the social architecture of respect and dignity. These new leaders will not have the loudest voices, but they will have the most attentive ears. Instead of pyramids, these postbureaucratic organizations will be structures built of energy and ideas, led by people who find their joy in the task at hand, while embracing one another—and not worrying about leaving monuments behind.

——

如果你去一家公司,你會問的最重要的問題是什麼?

假設有一個刻度表,從 1 到 10,10 代表 100%,1 代表與 0 相差無幾。你在你的工作上,投入了多少才賦?為什麼?

If you go into a company, what's the most important question you ask?

On a scale from 1 to 10, with 10 meaning 100 percent and 1 meaning close to zero, how much of your talents are being deployed in your job? And why?

――

▶ 奇葩與怪傑

羅伯・湯瑪斯（Robert Thomas）與班尼斯，在對近 50 位領導人進行訪談之後，於 2003 年發表了《奇葩與怪傑：時代、價值觀和關鍵時刻如何塑造領袖》（*Geeks and Geezers: How Era, Values, and Defining Moments Shape Leaders*）一書。書中所謂的「奇葩」（geeks），指的是年在 21 與 35 歲之間的領導人；所謂「怪傑」（geezers）指的是年在 70 與 93 歲之間的領導人。儘管年齡有這麼大的差異，這兩個領導族群有一些重要的共同點。

這項研究促成一種領導發展模式，成功的領導人有 4 種基本能力：能透過共同意義與他人打交道、有獨特而引人注目的聲音、正直，以及隨機應變的能力。在這 4 種能力中，又以應變能力最為重要。

湯瑪斯在接受我們訪談時，說明這本書的源起：

我的一個學生來找我，告訴我一家網際網路新創公司要他去上班，給他的待遇非常誘人，但他不知道究竟該怎麼做。他目前已有一份非常好的工作，那是一家名列《財星》雜誌 50 大排行（Fortune 50）的大公司，而且公司執行長對他很是賞識。他知道自己在這家公司做下去有非常好的機會與前途，但在若干方面，他又覺得如果繼續做下去，始終只能按照他人的劇本行事。

我就這個議題追問他，為什麼根據自己的主張行事對他這麼重要。他回答：「我發現我在把自己置於險境時，特別能學到許多東西。我想複製的是那種經驗。」

我說：「隨便舉個例子，說明你在什麼時候學到一些東西。」他說，他在負責一座塑膠廠的時候，有一次一名員工用了一條速成捷徑，結果造成安全問題，一條管線發生爆炸，這名員工也因傷重而終告不治。

但我這名學生說，他從這次事件學到一個教訓：身為領導人，不僅要完成上級交付的工作，不僅要想出新點子、做好舊有的那些事，還得像帶一個社群一樣進行領導。他告訴我，他從沒想到一家企業能像一個社群一

樣，他也從沒想到他會這麼關心他領導的員工。他擔任領導職位，是因為他在大學及其他地方受過訓練，對數字非常在行。

這次慘痛的經驗讓他震撼不已，但這件事的重點，不只是他現在知道自己應該為公司這個社群負責而已。為社群負責的覺悟，自然是一個重要教訓，但這次經驗還讓他發現一件事：當重大事件發生時，當你已經被逼到牆腳、退無可退時，或當你面對從未面對的狀況時，你會學到對你最重要的教訓。

這名學生對我說：「我不是那種喜歡找刺激、喜歡冒險的人。我不會讓自己陷入不必要的風險，不過我確實知道，只有在那樣的環境中，我才能學到大教訓。」

再回到他該不該轉往那家新創公司工作的問題。他說，在接到新創公司這項邀約時，他直覺認定這會是再一次學得大教訓的機會，他的寒毛都似乎豎了起來。

這位年輕人的經驗，讓我產生了一個構想：一個人在某些時間、某些地點學到的東西，會比在其他時間、其他地點學到的東西更重要。如果你善於覺察，發現這些時間或地點正在接近，或發現你已經置身其間，你應該可以學到很有價值的新東西。

所以，當班尼斯與我合寫這本書的時候，我們不僅強調「大考驗」概念的重要性，也不僅強調「大考驗」能助人學習，我們還要指出「大考驗」的另一種比較含蓄的意義：它為人帶來的不是一般的教訓。

每個人都能學到一些有關領導、有關做為領導人的教訓。這些教訓當然也都重要，但你若要他們加以描述，他們的說法往往了無新意，說來說去總脫不了教科書上那一套。真正讓我迷戀的，是另一類型的教訓。這類型的教訓能讓我們學到怎麼學。如果你能了解怎麼做才能學到重要的事——不只是知道如何累積知識、如何精通一件工作，而且知道怎麼做才能進一步了解自己，進一步了解自己的潛能——你一定可以學得很快。

I had a student who called me up and told me that he had a very attractive offer to go to work for an Internet start-up and didn't quite know what to do about it. He was currently in a position that was very good. It was a Fortune 50 company, and he was being mentored by the CEO. He had enormous opportunities and possibilities in front of him, and yet he felt that in some respects, he was going to be living out someone else's script.

And I pressed him about that issue, why writing his own script was so important to him, and his response was, "Well, I've come to recognize that I learn things when I put myself in danger, and it's that kind of experience that I want to reproduce."

I said, "Give me an example of a time when you learned something." He told a story of a time when he was responsible for a plastics plant and someone had done something that was a shortcut. He had created a very unsafe condition; a pipe blew up, and the person was scalded horribly and ultimately died from his injuries.

But my student said that he learned that being a leader is not just about accomplishing the objectives that are set out for you or coming up with creative ways in which to do old things, but somehow it's being a leader of a community. And he had never thought that a business would be a community or that he would ever care that much about the people he was leading because he'd learned through college and elsewhere to be really good with the numbers.

It took that experience to jar him, but the point wasn't simply that he now had to be responsible as a leader of a community. That's an important lesson to learn, but he also learned something else, which is that those kinds of big events, where your back is against the wall or you're encountering something that you've never seen before, are times in which you learn your most important lessons.

He said, "I'm not a thrill seeker. I don't put myself unnecessarily in danger, but I do know that I will learn big things only in those kinds of circumstances."

Going back to the conundrum he had about whether to take a job at a start-up, he said it felt like another one of those opportunities, almost as if the hair on the back of his neck was standing up, telling him that this was an opportunity to learn important things.

I drew from this young man's experience the idea that there are times and places in which one learns much more important things than one does in others. If you get good at spotting them, seeing them approaching, or recognizing that you're in them, you may, in fact, learn valuable new things.

So when Bennis and I did the book, not only did we emphasize the importance of the notion of crucibles and people learning from those kinds of experiences, but there was something else implicit in it, which was to distinguish between the types of lessons that people learned.

There were the lessons that they learned about leadership and about themselves as leaders. Those are important, but when people described them, they were often pretty mundane, the sort of thing that any textbook would capture. It was the second type of lesson that really intrigued me, which is the lessons that they learned about how they learn. If you gain some insight into what it takes for you to learn important things—not just what it takes for you to accumulate knowledge or what it takes for you to become versed in something, but what it takes for you to gain new insight into both who you are and what you could be—then that has to be an accelerant to learning.

——

▶ 怪傑有話要說

對好學不倦的班尼斯而言，與年輕一代溝通顯然是一個學習的好機會，我們也問他有關怪傑的問題。

怪傑們的人生歷練，不會比奇葩的更有趣嗎？

我很不願意這麼說，因為我也是怪傑其中一員。我很不願意語帶偏見或裁判，但怪傑活得久得多，經歷的事也比奇葩多得多。我認為，奇葩沒有經歷過像二次大戰或大蕭條那樣的大考驗。他們在成長、轉變的過程中，看

到的幾乎是連續不斷的繁榮、成長與成功。他們往往像是不知人間疾苦的富家子。

對他們的世界觀而言，911 事件是第一次集體震撼，是對他們的第一場大撼動。

Aren't the geezers inevitably more interesting than the geeks?

I would hate to say that, as I am one of them. I would hate to sound biased or judgmental, but they have lived longer and gone through an awful lot. I think what the geeks haven't experienced are the crucibles like World War II and the Depression. During their formative years, they have seen almost uninterrupted prosperity, growth, and success. They are often children of affluence.

September 11 was the first collective shock to the worldview they grew up with. It was a jolt to them.

———

你曾說，奇葩因面對的可能性太多而窒息。

是的，我認為可能性太多，確實令人焦躁不安。整個世界都是他們的，他們想做什麼就可以做什麼。他們擁有太多選項、太多可能性。

我每年都會開一個班，教領導的藝術與冒險，對象都是
20 來歲的學生。他們都是新一代中的佼佼者，卻因為對
前途一片茫然，最後都選擇進了法學院。依我看，這個
班的學生後來進法學院的，有三分之一的人是因為他們
除了進法學院以外不知道自己還想幹什麼。亞裔學生在
選擇科系的時候非常注重職業，因為他們是第一代美國
人。他們一般選擇理工與醫科，因為有了專業以後，更
能應付不確定。其他少數族裔也有這類現象，以猶太人
為例：「我那當醫生的兒子」，就是猶太人掛在嘴邊的
口頭禪。

You talk of the geeks being smothered in possibilities.

Yes. I think it does create anxiety. The world's their oyster, and
they can choose what they do. They have so many options and
possibilities.

Every year, I teach a class on the art and adventure of leadership
to a class of 20-year-olds. They are our best and brightest, but
they are so confused that they end up going to law school. I
think a third of those who take this class go on to law school
because they don't know what else to do. The Asian students are
highly vocationally oriented because they are first-generation
Americans. They tend to go into science or medicine. They

absorb uncertainty through the professions. You see that also with other groups—think of the Jewish expression "my son, the doctor."

———

但無論是奇葩或怪傑，似乎都對他們的世界觀非常肯定？

工作─生活平衡的議題，是不同世代之間的一大分歧。我曾問一個奇葩，如果這世上根本沒有電腦這樣的東西，他整天會做些什麼。這個奇葩想了半天，最後說「我不知道」，之後又加了一句：也許他會當一個悲慘的會計師。

But both the geeks and the geezers appear to be very certain of their view of the world?

The work-life balance issue is a real split between the generations. I asked one of the geeks what he would be doing if there wasn't such a thing as a computer. He paused for a long time, then said, "I don't know," and added that he would have been a miserable accountant.

———

介於奇葩與怪傑之間的那些人呢？那之間總不會是個無人地帶吧？

我曾經想過哪些人會讀這本書，依我之見，是年齡介於奇葩與怪傑之間的讀者，才會讀這本書。只有他們才會買這本書，因為奇葩不看書，而怪傑不會花錢買商管書（雖然我不認為這是一本商管書）。

與我合寫這本書的湯瑪斯，一直在旁邊嚷著：「那我呢？」介於奇葩與怪傑之間的人，不僅會買這本書，還是替這本書傳遞訊息的族群。我認為，這些中間族群有責任將這本書的訊息轉達給奇葩與怪傑。

我與我的女婿討論過這個問題，他40歲，有一大堆剛過40歲的友人，他們都為使用網際網路與科技而忙得不亦樂乎。他們都屬於這中間族群，既對科技並不陌生，年齡與智慧也比奇葩長了一些，傳遞訊息的工作非他們莫屬。

What about the people in between geeks and geezers? Isn't it a kind of no man's land?

When I think about the readership of the book, I think it is probably exactly that group in the middle. Those people should be the buyers. The geeks don't read, and the geezers don't buy books in the business category (although I don't think of this as a business book).

My coauthor kept saying, "What about me?" That group is not just the one that will buy this book, but it is an articulating point. These people have the responsibility, I think, to be the translators, the people who will help each group.

I was talking to my son-in-law, who is 40, and a lot of his friends who are just a little older are struggling with the Internet and the technology. So it is the people who are in the middle group, who are comfortable with the technology but are a little wiser and older, who have to be the articulating point.

———

你在與奇葩談話時，有沒有發現他們普遍比較重視自我認知？

我認為這個問題應該是這麼看的：奇葩比老一輩更認為自己有權討論自己、討論自己內心的感覺。這與一些怪傑很不一樣，他們連做夢都想不到竟能像這樣討論自己與家人的關係等。怪傑們很壓抑、很保守，年輕一代的人對感覺和抱負等，態度就開放得多。

Do you detect that same level of self-awareness in the geeks you talked to?

I think that what comes through is that they feel that they have more license to talk about themselves and their inner feelings. This is unlike some of the geezers, who would never dream about talking about their relations with their family and things like that. There is a real restraint among the geezers, a kind of reserve, while those in the younger generation are more free with their feelings, aspirations, and things like that.

———

談談我們培養領導人的方式？有許多人根本沒有自我認知的基礎，就進了企管碩士班。

美國有幾所大學已經有為期兩週的學前訓練，以團隊合作為訓練重點，還進行邁爾斯・布里格斯（Myers-Briggs）性格分類指標，所以學生都能有相當程度的自我認知。你必須了解一件事：商學院校大多數的教授，實際上沒有管理過任何組織，沒有實際領導的經驗。

許多商學院校現在已經招收工作了 3 到 5 年的學生，這一點我很高興。在許多事例上，這些學生的領導經驗比他們的教授還多。

我完全贊同建立一種全國性服務系統。我們迫切需要這種系統。年輕人都已經整裝待發，卻無處可去。我這麼

說，並不特指服務軍旅（但我不排除讓年輕人當兵），
我指的是給年輕人一條路，讓他們在進法學院或商學院
以前，先取得一些經驗。

哈佛商學院有一門必修的倫理課，大多數商學院校沒有
這樣的課。這是一個非常難的課題，但我們需要思考教
育的目的。我們必須在商學院校問這樣的問題：「這世
上有比金錢更重要的事嗎？公司之所以存在，為的不只
是賺錢、不只是獲利嗎？」有些事情當然比錢更重要，
公司當然不只為了賺錢，但我們必須解釋得更好。

**What about the way we develop leaders? A lot of people go into
MBA programs without a bedrock of self-awareness.**

Some universities in the United States have two weeks of
induction that concentrates on teamwork, and they do Myers-
Briggs so that the students become quite aware of who they are.
You've got to realize that most business school faculty members
have not actually run anything. They have not done the heavy
lifting of actually leading.

I am glad that business schools are now taking people who have
worked for three to five years. In many instances, these students
have more experience than the faculty.

I am totally in favor of a national service system. It is badly needed. The youth are all dressed up, but they have no place to go. This would not be specifically military service (although I wouldn't exclude the military), but a way for them to get experience before going to law school or business school.

There is a required course on ethics at Harvard Business School, but not at most business schools. It's a very difficult topic, but we need to think about the purpose of education. We have to ask the questions at business schools, "Is there something more important than money? Do corporations exist for something more than money and the bottom line?" Of course they do, but we have to explain it better.

———

對金錢的態度而言，奇葩與怪傑有差異嗎？

怪傑是在社會心理學家馬斯洛（Abraham Maslow）的求生模式中成長的。他們多數出身寒微，對財務的抱負也很有限。他們認為，一年能賺 1 萬美元就夠了。相形之下，一些奇葩在很年輕的時候，就已經賺了許多錢。這些人與他們那些老前輩生活在不同的世界裡，這些奇葩一旦破產，比較關心的會是怎麼生活，而不是身後留名的問題。

Was there a difference between the geeks and the geezers in terms of their attitudes toward money?

The geezers were brought up in Maslow's survival mode. Often they grew up in some poverty with limited financial aspirations. They thought that earning $10,000 a year would have been enough. Compare that to the geeks, some of whom made a lot of money when they were young. They are operating in a different context. If they were broke, they would be more concerned with making a living than with making history.

———

奇葩與怪傑的差距這麼大,該怎麼溝通?

我們必須想辦法溝通這差距。如果能聽到許多家庭開始討論這本書,事情就有趣多了。依我之見,怪傑或許比較不容易因應時下做出改變——如科技改變等。人一旦到了 60 幾歲,難免想到大限將至,對年輕一定會有某種豔羨。人到了 60 幾歲,一般也不會再有多少雄心壯志了。

世代之間的溝通對話很重要。奇異(General Electric)等幾家公司,已經展開一種逆向師承制,由年輕的人提攜年長的人,幫助他們熟稔電子世界。年齡歧視是個很普遍的問題,我已經 80 幾歲的這個事實,或許我對這類問題

早已見怪不怪。大家看到我開車的時候，當他們看到一頭白髮，行為也會變得不一樣。這些都是社會整體的深奧議題。

How can you bridge the gap between the geeks and the geezers?

We must. It would be interesting to hear families discussing the book. I think the geezers may have a more difficult time with the changes that are under way, such as technology. You start to think about your mortality when you are in your sixties, and there is a certain envy of youth. When you are in your sixties, you are no longer promising.

Dialogue between generations is important. A number of companies, including GE, have reverse mentoring, where young people mentor older people to acquaint them with the e-world. There is a lot of ageism, which I probably wouldn't be sensitive to except for the fact I am in my eighties. When people see me in a car, they see white hair, and they behave differently. These will be profound issues for society in general.

——

第 3 章

第五級領導

策略大師
C.K.普哈拉
《A 到 A$^+$》作者
與談人│**吉姆・柯林斯**

現代領導人往往與「謙卑」扯不上什麼關係；事實上，「謙卑」與「領導」幾乎已經成為一種相互矛盾的名詞，但其實不必這樣。

高舉「謙卑」大旗往領導這項知識大進擊的，是普哈拉（C.K. Prahalad）。他在 2007 年與 2009 年的「50 大商業思想家」排行名列榜首，不幸於 2010 年英年早逝。普哈拉在去世前不久與我們會面時，曾談到領導。

你認為，新一代領導人應該有什麼特質？

我認為謙卑是一個很好的出發點。我覺得，在我們今天的社會，太多人認為想當領導人就必須傲慢。這種想法是錯的，因為首先，領導的要旨在於希望，在於改變，在於未來。如果從這 3 個前提著眼，領導人必須願意聽取他人的意見，因為未來是難料的。讓人告訴你過去的事不難，因為過去的事已經塵埃落定。但未來的事沒那麼確定，你必須聆聽，蒐集各式各樣的意見及看法。

我打個比方，好的領導人像牧羊犬，而好的牧羊犬必須遵守 3 條規則。第一條，你怎麼叫都可以，但不能咬。第二條，你必須守在後面，不能跑在羊群前面。第三條，你必須知道往哪裡走，而且不能讓羊群走失。

我始終都這麼說，是因為我認為如果你仔細思考，就會發現領導就是構築共識。一旦有了股東，就必須擔心共創的問題；你必須聆聽，必須構築共識。你會聽到各式各樣的意見，那就像怎麼叫都可以，但不能咬一樣。要知道，那些另有看法、意見與你不同的人，可能比那些意見與你一致的人更有價值。有一句老話，不就這麼說：如果你自己也會叫，幹嘛還要養狗？如果你手下的人各個都同意你的看法，既然你已經知道答案，又何必用這麼多人？所以，想了解新領導人像什麼樣，就不能不談異議。

除了要能夠接受異議，你也必須對未來有看法。如果你對未來沒有明確看法，是無法勝任領導的。不過，大多數領導人沒有看法，或者他們雖然有看法，卻沒有清楚表明。為股東增加財富，並不是一種對未來的看法，只要把事情做對，自然就可以辦到這點。

所以我說，領導人除了必須對公司的願景有明確看法外，還要知道……社會的基本結構，要知道公司應該怎麼參與、怎麼塑造這個社會，要清楚了解如何構築共識、如何聽取異議。

如何從後方領導？這主要就是謙卑的問題了。

最後，我要說，未來的領導人要具備更多道德權威。這
不是階級架構帶來的權威。誰擁有巨大的道德權威？
想到這個問題，就讓我們想到甘地。甘地沒有大軍，他
有的只是道德力量。而且甘地還有一項長處：他絕不武
斷。他很強悍，雖曾多次獨斷獨行，但他願意改變他的
做法。

而且他也說，手段與目的一樣重要；不要在英國人遭遇
問題的時候乘人之危。我的意思是說，甘地有好幾次
可以乘機打擊英國，當時每個人都勸他採取行動。但他
說：「不行，這次不行；我們得等。因為手段與成果一
樣重要。」甘地擁有巨大的道德權威，大家都聽他的。

甘地不是那種事事講求民主的人，但他願意聽取很多人
的意見，而且有很明確的價值觀。我要談的最後一點，
就是這種價值觀。我喜歡講個故事來說明這點，如果有
人跑到甘地面前，對甘地說：「你既然已經宣布全面
『自治』（swaraj），我們已經全面自由，何不出去殺
1 萬個英國人？」甘地一定不會同意這麼做。他會說：
「我們不能用這種方式爭取自由，因為我們採取的做法
具有深度價值。」這種價值就是非暴力。印度必須用和
平抵抗的方式來爭取獨立，必須用和平手段來對抗不公
正的法律。

我想，這些就是我對領導人的想法。

What do you think will be the characteristics of the new generation of leaders?

I think humility is a good start. I think we got to a point where people thought that if you wanted to be a leader, you had to be arrogant. No. First, leadership is about hope, leadership is about change, and leadership is about the future. And if you start with those three premises, I want leaders who are willing to listen because the future is not clear. People can tell you about the past because there's certainty about the past. With the future, there's not much certainty, so you have to listen and bring in multiple perspectives.

Let me use a metaphor. I look at good leaders as sheepdogs. Good sheepdogs have to follow three rules. Number one, you can bark a lot, but you don't bite. Number two, you have to be behind; you cannot be ahead of the sheep. Number three, you must know where to go, and you mustn't lose the sheep.

I always say that because I think that if you think about leadership, it's about consensus building, because when you have stakeholders and you have to worry about co-creation, you must listen and you must build consensus. You can have multiple conversations, but it's the equivalent of barking a lot, but not biting. People who tell you things that are different from what you think may be more valuable than people who agree with you because, as the old saying goes, if you can bark yourself, why have a dog? If all the people you have agree with you, then why do you have so many people? You already know the answer. So dissent is an integral part of understanding what a new leader will look like.

And you must have a point of view about the future. You cannot lead unless you have a point of view. However, most leaders do not have a point of view, or if they have one, they don't express it clearly. Increasing shareholder wealth cannot be a point of view about the future. It's incidental to doing the right things.

So I would say that the leader needs to articulate a point of view on not only where the company can be, but . . . what are the underlying structures of society and how the company is going to participate in and shape that society, and to have a clear understanding of how to build consensus and listen to dissent.

How do you lead from behind? That means a lot of humility.

And finally, I would say that the leaders of the future will have more moral authority. This isn't hierarchical authority. People who come to mind who have tremendous moral authority are people like Gandhi. If you think about Gandhi, he did not have big armies. His force was moral force. And the virtue in Gandhi was that he was never dogmatic. He was tough. He was autocratic many times, but he was willing to change his methods.

And he also said that means are as important as ends; don't take advantage of the British when they have problems. I mean, he could have taken advantage many times. Everybody advised him to take advantage. He said, "No, not this time; we will wait, because means are as important as outcomes." So he had tremendous moral authority. People listened to him.

Now, he was not always the most democratic person, but he listened to a lot of people and he had clear values, which I think is the last thing I want to say. My favorite thought experiment is that if somebody had gone to Gandhi and said, "You declared full swaraj, total freedom; why don't we just go and kill 10,000 Brits?" that would have been no go. He would have said, "That is not how we are going to get freedom because there are deep values involved in how we are going to do it." It was nonviolence. It was peaceful resistance, fighting against unjust laws through peaceful means.

And so I think that would be my view of thinking about a leader.

——

我們記得，你之前曾告訴我們，有時你會將甘地與巴頓（George S. Patton）將軍做對比。巴頓的領導類型與甘地非常不同。巴頓式的領導有什麼優點嗎？你怎麼看？

巴頓式的領導有許多優勢。但你想一想，他運用階級地位展示聲望、權威。沒錯，他是一位非常成功的將領，但如果你看過那部電影，應該還記得巨幅美國國旗迎風飄揚、官校學生戎裝整肅的那一幕。巴頓對他們說，有一天，你含飴弄孫，小孫子問你：「爺爺，你在大戰的時候做什麼？」你不用面帶慚色地說：「我在某個地方鏟糞堆肥。」

巴頓說這話，到底是什麼意思？他要說的是：「我向你們保證，你們會活下去，不要怕。」那些 19、20 歲的年輕人在擔心什麼？他們擔心的是戰死沙場，事實上這也是當時許多年輕人的下場。但巴頓對他們說：「不要怕，我們會贏。」而且他從不使用「德國」（German）這個字詞。他只說：「我們要殺『蠻子』（Hun）。」這麼做的原因很簡單：因為當年美軍裡面，有很多德裔美國人。

巴頓的領導力也很強，但甘地與巴頓的領導差異在於，甘地用的方法更能持久。甘地說，如果我們採取以眼還眼的手段，整個村子的人都會變成瞎子。這是一種非常不一樣的思考方式。不要以武力對抗武力，要以道德權

威，當然還要以邏輯思考、以構築共識來對抗武力。甘地用的是一種另類的衝突解決之道。

我相信，有時候我們確實需要像喬治‧巴頓這樣的領導人，但在許多時候，透過共識同樣可以解決衝突。採用這種做法需要了解、同理心、時間與耐性。所以，如果你志在謀求世界和平與安定，依我看，以武力對抗武力絕對不是好辦法。

如果仔細思考，你會發現以對話與討論的方式和平解決問題，成功的可能性反而比較大。事實上，我不知道這話是誰說的，但無論怎麼說，這話很值得我們深思：你不必與友人談和；只要與敵人談和就好了。換句話說，只有在其他人與你意見不合的情況下，你才需要不斷地談。如果其他人都同意你的看法，何必還要多費口舌？

We remember you telling us before that you sometimes contrast Gandhi with General Patton, who was a very different type of leader. Are there any virtues to the Patton style of leadership? How would you characterize that?

Patton's style of leadership had many advantages. But think about it; hc was using hicrarchical position as a way of demonstrating credibility. Now, he was a very successful general, but if you see the movie, which is always fascinating for me, big American flag and all beautifully dressed up and telling young kids, when you are rocking your grandchild on your knee—I'm paraphrasing—and he asks you, "Grandpa, what did you do during the Great War?" you don't have to say, "I was shoveling shit in some place."

Essentially, what is he saying? I assure you, you will live, don't be afraid. Because what is the worry that a 19-year-old has, a 20-year-old? That he's going to get shot, which many of them did. But he is telling them, "Don't worry, we will win." And he never once used the word German. He said, "We'll kill the Huns." And the reason is very simple: there were lots of Germans or soldiers of German extraction in the army.

So Patton was also powerful, but the difference between Gandhi and him is that Gandhi's methods will endure longer. Gandhi said that if we take an eye for an eye, the whole village will be blind. It's a very different way of thinking about it. You don't fight force with force; you fight force with moral authority, and you certainly fight force with logic and consensus building. So it's a different way of resolving conflicts.

I do believe that sometimes we need George Patton, but many times consensus also means conflict resolution. It requires understanding, empathy, time, and patience. So if you think about global peace and stability, it's not at all clear to me that force for force is the right way to go.

So in a funny way, I think that if you start thinking about it, coming to terms peacefully through dialogue and discussion is a safer bet. In fact, I don't know who said this, but anyway, it is an interesting thing. You don't have to make peace with your friends; you only have to make peace with your enemies. So in other words, where people don't agree with you is where you need to talk a lot. Why keep talking to people who all agree with you?

——

▶ 基業長青

　　柯林斯見過非常多領導人，為了了解為什麼有些公司歷久不衰，有些公司倒在路邊，有些公司還不錯，有些公司好得出奇，有些公司卻一敗塗地，他觀察了許多公司，並且根據這項研究寫了幾本書。其中，最有名的包括與傑利・薄樂斯（Jerry Porras）合著的《基業長青》、《為什麼 A⁺ 巨人也會倒下》（*How the Mighty Fall*）、《從 A 到 A⁺》，以及與莫頓・韓森（Morten Hansen）合著的《十倍勝，絕不單靠運氣》（*Great by Choice*），還有其他無數篇論文。

　　《基業長青》在 1994 年出版，這本雄踞美國《商業周刊》（*Businessweek*）暢銷書榜長達 6 年多的書，主要探討的是所謂「百年企業」（visionary companies）的必備特質。歷久不衰的公司一定有個有效的願景，能將公司的核心意識型態具體化。這種有效的願景又有兩項要件：核心價值（一種指導原則與信條系統）與核心宗旨（組織存在的最基本理由）。

　　柯林斯說，核心價值「只有少數幾條指導原則，是組織最重要、最持久的信條；這些原則既不能與特定文化或營運做法混為一談，也不能因財務利益或短期權宜之計而稍有妥協」。他與薄樂斯寫道：「歷久不衰的公司，都有核心價值與一套核心宗旨，這些公司的商業策略與做法，儘管會因應不斷變化的世界而持續調整，他們的核心價值與核心宗旨卻永遠不變。」

　　在《基業長青》出版以前，柯林斯在史丹佛大學商學院（Stanford Graduate School of Business）從事研究與執教。他在史丹佛教了 7 年，之後回到科羅拉多州巨石城（Boulder）老家，創辦一家管理實驗室，與公、民營企業及社會事業高階主管合作展開多項長期研究計劃。在這個實驗室裡，柯林斯成了「一位自雇的教授，自己出錢捐助

自己的講座，自己給自己職權」。他的實驗室從統計學觀點，檢驗商務議題與結構。他說：「其他人喜歡意見及看法，我則偏好數據資料。」

▶ 從 A 到 A⁺

1996 年，柯林斯與研究團隊將注意力轉移到另一個極有挑戰性的商務議題：一家好的公司，能不能更上層樓，成為一家偉大的公司？如果能，要怎樣才能做到？柯林斯認為，如果能找出哪些公司能突破界線、由好而成就偉大，哪些不能突破，他就可以比對、過濾，找出從 A 晉升到 A⁺ 的要件。

經過 5 年研究，柯林斯得到一些答案。他的研究團隊找到幾項使好公司變成偉大公司的要件，但柯林斯指出，其中最讓人稱奇的，就是領導類型的要件。經過抽絲剝繭的過濾，好公司與偉大的公司，差別似乎就在於領導類型的不同。或許柯林斯在一開始也不相信只憑一個人的領導型態，就能造成這麼大的差別（好公司與偉人公司之差），但他的研究數據透露的並不一樣。

　　柯林斯提出一套領導學新術語，來描述這一類領導人：「第五級領導」。他以領導技巧為準，將領導分為五級。第一級領導為個人能力，領導人運用個人知識與才能為組織貢獻。第二級領導為團隊技巧，以及如何在一個團體內有效領導。第三級領導，領導人展現管理能力，能讓其他人組織起來，為共同目標努力。第四級領導是非柯林斯式的傳統領導：領導人能表明一種願景，激勵他人表現。第五級領導是領導的最高境界。第五級領導人是偉人，第四級領導人只是優秀而已。

　　第五級領導人除了擁有前四級的一切領導技巧之外，還有一項特質：他們能為了組織，放下追求自我與自利的需求。第五級領導人對公司及公司的使命，有一種幾近英雄式的承諾。他們將一切情感完全投入公司，既無心於自我提升，也沒有餘裕這麼做。他們的優先事項總是公司，但這類領導人絕不是單槍匹馬，他們需要一支好團隊，如何創建這樣的團隊也是他們的責任。

　　柯林斯指出，公司在聘請高階主管時，應該物色的是第五級領導人。但這類領導人長得像什麼樣子，要從哪裡找？想物色這樣的領導人，首先需要注意的是，他們不是那種飛揚跋扈、自我中心的名流。許多人原本以為，想將

公司從「好」轉型為「偉大」，就得靠大名鼎鼎、魅力十足的領導人。第五級領導理論並不同意這項假定，在柯林斯這項 5 年研究中脫穎而出的最高階領導人，相對而言，在他們的產業之外都不甚有名。這項研究成果似乎也指出，領導重心已經從英雄式轉為非英雄式。仔細觀察這類組織，你會發現它們的表現儘管卓越，卻沒有人將一切功勞攬在自己身上。

根據柯林斯的研究，謙卑是第五級領導的關鍵要件。他提出一個簡單的方程式：謙卑＋意志＝第五級領導。柯林斯指出：「第五級領導人是一種二元性的研究，他們既謙和又剛愎，既害羞又無畏。」他說，這類領導人有幾種特性，儘管促成組織的偉大成就，卻從不妄自尊大，寧可置身幕後。在達成組織目標的過程中，他們堅定不移。但他們之所以能夠激勵組織奮發而前，靠的並不是個人魅力，而是原則與標準。他們為組織的永續經營、常勝不衰進行規劃，他們勇於甚至盼望為組織建立有效接班。

一旦事情出現差錯，第五級領導人非但不會推諉責任，還會樂於一肩承擔。當一切進展順利的時候，他們總是忙著讚美他人，忙著稱許團隊的貢獻。只要你能改變對領導力的省思，採納第五級領導的觀點，組織的人才地

圖會立即出現變化。在大多數組織內都能找到第五級領導人，但問題是英雄式高階主管的信念，早已充塞商界每一個角落。

柯林斯說：「如果我們坐視不管，讓搖滾樂明星一樣的名流，繼續在商界呼風喚雨，各式各樣的企業與組織終將由盛而衰。20 世紀是偉大的世紀，但到了 21 世紀，能夠長盛不衰的組織將少之又少，這是我們非常可能面對的情況。如果真如我的想像，『好』是『偉大』的敵人，目前的領導趨勢正將決定性優勢供手讓給敵人。」

▶ 十倍勝

現代世界是個一團混亂的地方，面對這種種亂象能因應自如的領導人不多。在出版《從 A 到 A⁺》以後，柯林斯卯上另一個讓他深感興趣的問題：有些組織與領導人特別能在混亂中取勝，這是怎麼回事？柯林斯用他的老方法解決這個問題：他先找出善於應付亂局、能在動盪不安中取勝的公司，再找出不善因應混亂的組織，然後進行檢驗，觀察兩者之間有什麼不同。

一開始，柯林斯與團隊蒐集了 20,400 家公司，經過逐步篩選，最後剩下 7 家公司。柯林斯團隊稱這 7 個組織為十倍勝（10X）公司，因為它們的表現至少比業界指數強 10 倍。為什麼這幾個組織能在混亂不堪、變化迅速的環境中勝出，其他公司卻辦不到？柯林斯說，造成這種表現差距的一項主因，就是領導。

柯林斯用爭著成為史上第一組抵達南極的勞德・艾蒙森（Roald Amundsen）與羅伯特・法爾肯・史考特（Robert Falcon Scott）團隊故事為例，來說明這項差異。這兩支南極探險隊擁有的資源與能力都差不多，也都在同樣艱難的環境下運作。但最後，艾蒙森的探險隊成功抵達南極，史考特的探險隊在回程途中困於冰雪，全隊死難。造成這種成敗、生死之差的主因，就在於行為差距。而這種行為差距，很像十倍勝公司領導人與其他遇亂則表現不佳公司領導人之間的差距。

十倍勝公司領導人，在面對一個他們大體上難以控制的世界時，會想辦法控制他們還能控制的一些層面。柯林斯說，這類領導人「擁抱一種既可以控制、又難以控制的矛盾」。他們了解也接受世上有些事變幻莫測、無法控制的事實，但他們不接受生死有命、成敗在天的想法。

　　柯林斯說，大體而言，十倍勝公司領導人展現 3 種行為特質：「狂熱性的紀律、實驗性的創意，以及生產性的偏執。」或許，在柯林斯這套十倍勝領導人理論中，最有趣的，是他所謂「20 哩行軍」（20-Mile Marching）的概念。根據這項概念，亂中取勝的祕訣，就在於在一段長時間中，一步一步不斷地往前進。即使有可能超越「20 哩」，無論它代表的目標是什麼，十倍勝領導人會按兵不動。但另一方面，無論情勢多麼艱困，十倍勝領導人總會想盡一切辦法，達到最接近那個概念性的「20 哩」里程碑的目標。十倍勝領導人很有抱負，但他們懂得自律與自控，知道什麼時候不應該躁進。

　　以前述艾蒙森的個案為例，他的探險隊步步為營，向目標前進，即使在天候狀況良好時也不躁進。柯林斯說，相形之下，史考特探險隊的進度卻是走走停停，天氣好的時候多走一些，天氣壞的時候裹足不前。

▶ 第五級領導

　　在與柯林斯的訪談中，我們一開始就提出第五級領導的問題，他很快就提出一個較廣泛的看法。

能與我們談一談第五級領導嗎?

首先,我要說明一下。概括來說,我不是很贊同以執行長為中心的領導觀。我發現,世人有關領導的解說過於簡化,經常犯下錯誤將太多變數一語帶過。一家公司的表現如果很好,我們就說那是因為公司的領導很高明;如果表現不好,我們就說它的領導不如我們想像那麼高明。

我認為,我們目前對領導的觀念,仍處於黑暗時代。古早以前,每當發生地震、穀物歉收或疾病而我們不明究理的時候,我們就把原因推給神。之後,文藝復興與啟蒙時代來臨,我們在物理與化學領域有了新發現,於是能對地震與穀物歉收提出一些與過去不同的解釋。

在 20 與 21 世紀,面對人為社會世界的一切總總,我們仍處於黑暗時代。我們動不動就將問題歸於領導,這就是明證。今天,世人對領導的觀念,就像古早世人對神的觀念一樣。我的意思並不是說,你必須是無神論者才行,但如果碰上問題,能夠不單只從神或從領導這兩方面尋找答案,就會找到其他很重要的因素。

What can you tell us about *level 5 leadership*?

First, let me back up a bit. In general, I have a bias against a CEO-centric view of the world. Leadership answers often strike me as oversimplistic and in danger of covering up too many variables. If a company does really well, we say that the reason was great leadership; if it doesn't do well, we say that the leadership wasn't as great as we thought.

I see leadership as being in our version of the Dark Ages. In an earlier period, whenever we didn't understand something—an earthquake or crops failing or disease—we would ascribe it to God. But then came the Renaissance and the Enlightenment and we discovered new areas of physics and chemistry, so we could offer different explanations for earthquakes and crop failures.

In the twentieth and twenty-first centuries, when we're looking at the social world, the man-made world, we are still in the Dark Ages. This is shown by our predilection for looking for leadership answers. Leadership is to the twentieth century what God was to a much earlier period. That doesn't mean that you have to become an atheist. But if you stop looking for answers that are always either God or leadership, you will find other underlying factors.

———

這個領導觀點對你的研究有什麼影響？

我在研究過程中，總是告訴研究人員，先不要考慮領導

人的角色，這樣我們才能找到其他要素。我們假設還有其他可以發掘的東西——物理定律。循著這條軌跡走下去，我不指望能發現《從 A 到 A⁺》那本書裡出現的第五級領導概念，我甚至不想發現這種概念，那不是我們要找的東西。

在下標的時候，我很不喜歡章標出現「領導力」這個字詞。對我來說，這樣的標題幾乎代表失敗；當然，事實上並非如此。我的想法是，希望能找到一些有趣的東西，那是一件美事。

研究團隊表示，我們認為這些公司的執行長，對公司能不能跨入另一級、能不能從好公司轉型為偉大公司影響非常大。我答覆他們說，規模差不多的對照組公司也有領導人，而且其中還有很多非常優秀的領導人，但表現卻不好。所以，不能說問題出在領導，因為這兩組公司都有傑出領導人。也因此，領導不是變數，我看你們還得從頭來過，另外找出一些有用的東西。

但研究團隊不服氣，並對我說：「吉姆，不瞞你說，我們認為你搞錯一件事：你把領導看成一種二元化的概念，你認為，一位領導人要不偉大、要不不偉大，這就是問題的關鍵。但我們的看法是，領導問題比這精微、

奧妙得多了。」基於這項論點，我們產生了一個構想：
領導是不斷演變的一套能力與成熟度。所以，問題的重
點，不在於領導人好或不好，而在於「處於哪個領導階
段」和成熟度有多少。

我們根據這個構想，歸納出一個看法：表現最優、而且
能夠歷久不衰的這些公司，擁有第五級領導的特質，而
另一組公司有的只是第四級領導人。所以，重點不在於
領導，真正的重點是，你是第四級領導人、第三級領導
人，還是一位第五級領導人？就這樣，我們引伸出一個
問題：「第五級領導人的特質是什麼？」為什麼能從眾
多領導人中脫穎而出？

So how has this view of leadership influenced your research?

In our research, I've always said, let's discount the role of the
leader so that we can find the other factors. Let's assume that
there are other things to discover—laws of physics. So going
down that path, the *level 5 leadership* finding that came out in
Good to Great was not what I expected to find. I didn't even want
to find it. This was not something that we were looking for.

I was very uncomfortable with having a chapter title with the
word leadership in it. To me this almost felt like a failure, which
of course it wasn't. It was good that we had found something

interesting.

What the research team said was, we think that the CEOs of these companies had a huge impact on whether they shifted from one level to another, whether they went from the good level to the great level. My reply was that the comparison companies also had leaders, exceptional leaders in many cases, but the companies ended up not performing as well. So you can't say that the answer is leadership, because you have outstanding leaders in both sets of companies. Thus, leadership is not a variable, so go back and do something useful and look for other stuff.

But the research team pushed back at me. They said, "Jim, what we really think you're missing is that you're looking at this as a binary idea—either you are a great leader or you are not, and that's the critical question. But what we believe is that it's much more nuanced than that." That eventually led to the idea that leadership is an evolving series of capabilities and levels of maturity. So it's not a leadership or not question, it's a "what stage of leadership" question and what level of maturity you have reached.

In turn, this led to the insight that those companies that tend to produce the best, most sustained results over time have the characteristics that we put at level 5, and other companies tend to have leaders who are stuck at level 4. So the issue wasn't leadership; the real issue was, are you a level 4 leader, a level 3 leader, or a level 5 leader? And that then led to the question, "What are the characteristics of the level 5 leader?" What cloth were they cut from that made them different from the others?

———

你的結論是什麼？

最後，我們總結出一項基本定義。自 1996 年展開這項研究以來，我不斷深入思考，愈想愈覺得這項定義有道理。第五級領導的核心，是一位胸懷大志、全心全意、以公司大業與工作為第一優先的領導人。這位領導人為實現自己的抱負，有鋼鐵般的驚人意志。會認為自己只是公司的一部分，為的不是自己，而是公司與公司的長遠利益。這種領導人不是那種乖乖牌，不是那種軟弱、像牆頭草的領導人，事實上正好相反。

And what were your conclusions?

That it came down to one essential definition. And the more I live with this definition since the research, which started in 1996, the more comfortable I am with it. The central dimension for level 5 is a leader who is ambitious first and foremost for the cause—for the company and for the work, not for himself or herself—and has an absolutely terrifying iron will to make good on that ambition. It is that combination—the fact that it's not about the leader, it's not first and foremost for him or her, it's for the company and its long-term interests, of which the leader is just a part. But it's not a meekness; it's not a weakness; it's not a wallflower type. It's the other side of the coin.

———

以公司為第一的領導人？

就是那種只要能讓公司更好，不惜開除自己親兄弟的領導人。這樣的領導人會把自己的身家性命賭在公司的前途上。只要能讓公司更好，即使冒著生命危險，他們也會在最危急的關頭為公司情義相挺。只要能讓公司更好，他們甚至願意拱手讓出執行長的大位。他們願意為公司付出一切。不論多痛苦，不論多煎熬，只要為了公司，他們會做一切必要的決定。第五級領導人能從眾多領導人中脫穎而出，就是因為具備這些特質。

Someone who puts the company first?

People who will fire their brother if that is what it takes to make the company great. They will bet the company. They will put their own lives through the worst circumstances, if that is what it takes to make the company great. They will even step away from the CEO role, if that is what it takes to make the company great. They will do whatever it takes. No matter how painful, no matter how emotionally stressing the decision may be, they have the will to do it. It is that very unusual combination that separates out the level 5 leaders.

———

你打算將研究擴展到非營利事業與社會領域嗎？

也好，就讓我來預告一下我的工作，我對非商業領域愈來愈感興趣。今天早上，我就在撰寫一篇有關社會領域「從 A 到 A⁺」的論文。我之後會針對非營利醫療組織，運用同樣的配對模式——其中一組組織能由「好」而「偉大」，另一組不能——展開研究。

我相信，到最後，我們的研究做出的最大貢獻，就在於我們的方法。事實上，我們用的這套方法，只是從研究方法演繹而來，並非我們首創。最讓我感興趣的問題是，怎麼營造一個偉大的社會？想營造一個偉大的社會，首先你必須讓所有社會組成分子都偉大。

如果我們的社會有的只是一些偉大的公司，雖然我們或許也會有一個繁榮的社會，但不會有一個偉大的社會。

想建立一個偉大的社會，只有一些偉大的公司還嫌不夠；我們還必須有偉大的學校、有偉大的收容救濟設施、有偉大的交響樂團演奏偉大的音樂，還必須有偉大的醫療系統。此外，我們還必須有偉大的政府、偉大的政府機構、偉大的城市，以及偉大的警察局。這些都是營造一個偉大社會的基石，我個人覺得這個問題讓我充滿鬥志。

我現在做的還是商業研究，但這或許是我做的最後一個商業案了。在探討社會領域「從 A 到 A⁺」的問題時，商界與社會領域之間的差異讓我稱奇不已。我絕對相信這項構想可以運用到社會，甚至更廣的領域中。

如果我要強調一件事，我要強調的是商界與社會或政府領域之間的重大差距。許多人說，想解決非商業領域的問題，最主要的解決之道，就是讓非商業領域更像商界。這是一個大錯。這個答案不對。因為我們現在知道，大多數企業的作為與成果扯不上關係。此外，既然大多數企業的成績只是平庸而已，為什麼要引用一些不怎麼好的做法？所以，真正重要的差距不在商界與社會之分，而在「好」與「偉大」之分。

Do you intend to broaden your research to the nonprofit and social sector?

Well, as a little preview of my work, I am increasingly interested in the nonbusiness sector. This morning I was working on a big article I'm writing on good to great for the social sector. Eventually I want to do research in which we use the same matched-pair method and apply it to nonprofit health organizations—one that went from good to great and one that didn't.

Ultimately, I believe the greatest contribution of our work is our method. It's actually not our findings. Our findings are just derived from the research method. The question that is of most interest to me is, how do you build a great society? You do it by ensuring that all the component parts of society become great.

If we had a society that only has great companies, we'd have a prosperous society, but we wouldn't have a great society. To be a great society, it isn't enough to have great companies; we must also have great schools and great homeless shelters, and we have to have great orchestras that play great music, and we have to have great healthcare systems. Ultimately, we have to have great governments, great government agencies, great cities, and great police departments. These are all building blocks of a great society. Personally, it's a question that I find very energizing.

I'm doing a piece of business research now, but it might be the last piece of business research I do. As I work on the good to great social-sector question, I'm really puzzling over the differences between business and the social sector. I'm absolutely convinced that the ideas also apply to the social sector, and in some ways apply even more.

If I can highlight one thing, it's the critical distinction between the business sector and the social or government sector. A big mistake people make is to say that the primary solution to the problems of the nonbusiness sector is for the nonbusiness sector to become more like the business sector. That's the wrong answer. Because what most businesses do, we now know, doesn't correlate with results. And since most businesses are mediocre, why would you want to import the practices of mediocrity? So the real distinction is not between business and social; it's between great and good.

———

這麼說，你的構想同樣適用於公共組織？

完全正確。在公共組織的環境中，第五級領導的應用面，甚至可能更廣。而且，在一個像英國國家衛生事業局（National Health Service）這樣的組織中，想做到第五級領導幾乎可以說比較容易，因為在這裡工作的人懷抱熱情，為的是理念而不是金錢。

我還要說一件事，而且這件事事關重大。你不妨自問，這麼多轉入非商業性事業領域的商界主管，最後以失敗收場，為什麼？部分原因是，在社會性較濃的非商業性組織裡，當一個第五級領導人需要一種更像是議員的領導風格。你主要是一位第五級領導議員，而不只是一位第五級領導主管。如果你把執行長那套主管風格，搬進權力比公司分散、也比公司複雜得多的議會，最後的輸家會是你，贏家是權力。能夠了解這個道理、將領導風格從商界主管模式轉型為議員模式的主管，儘管仍是第五級領導人，一般比較容易成功。

也因此，領導沒有準確的定論；你如果還像《從 A 到 A$^+$》那些執行長一樣，事情未必能做得好。你當然要當第五級領導人，不過是一位議員模式的第五級領導人。

So your ideas are just as relevant for public organizations?

Exactly. Level 5 applies even more in that sort of environment. And it's almost easier to be level 5 in an organization like the U.K.'s National Health Service because of the passion, because of the cause—people certainly aren't doing it for the money.

There's one thing that I would say, and this is really key. Ask yourself, why do so many business executives who move into non-business-related sectors fail? And they often do. In part, it is because being a level 5 leader in these more social institutions requires a more legislative style of leadership. You are a legislative level 5 more than an executive level 5. If you bring a CEO executive style into a legislative setting, where the power is much more diffuse and complex, you will lose. The power will win. Those executives that understand that they need to shift from the executive model to the legislative model, while still being level 5, tend to have an easier time.

So, then, an exact manifestation would not apply; you wouldn't want to be like the CEOs from *Good to Great*. You want to be a level 5, but you want to be a legislative level 5.

——

《基業長青》熱賣了幾乎 10 年。為什麼？

看到這本書不斷熱賣，我也一直很吃驚。這本書在暢銷書排行榜上前後 64 個月，我想我們賣了 100 萬本。為什麼這個話題一直能讓讀者感到興趣？我猜，這其間有 3 個原因。首先，傑利‧薄樂斯與我在書中討論的，都是 20 世紀的標竿型企業，我們以 IBM、索尼（Sony）與華特‧迪士尼這類公司為重點，這吸引了一大堆讀者。其次，我們的研究品質顯然經得起時間考驗；這本書以獨創一格的方式，從歷史（回溯公司的起源）與對比（與主要

競爭對手做比較）兩個角度觀察公司。最後，這本書談到許多人性面的事；我們不在意以非商業性的人性議題與商業議題混在一起，一併討論。

Built to Last has been a big seller for almost a decade. Why?

I'm continuously surprised to see that the book keeps selling and selling. I think we're pushing 1,000,000 copies in print after some 64 months on the bestseller lists. Why does the topic continue to fascinate readers? My guess is that there are three reasons. First, Jerry and I talked about the corporate icons of the twentieth century; we focused on companies like IBM and Sony and Walt Disney. That draws a lot of people. Second, the quality of our research has obviously stood the test of time; the book uniquely looked at companies both historically (going back to their roots) and comparatively (against their major competitors). Finally, a lot of what's in that book is revelations about humans at work; we weren't afraid to have business findings mixed in with nonbusiness, human findings.

———

我們怎麼在一個組織裡，辨認你所謂的「核心價值」？

首先，一個組織不需要有明確的核心價值。核心價值不必冠冕堂皇，甚至可以兇狠殘酷都沒有關係。而且，也

不必講求人道，不過在大多數個案中，公司的核心價值都很重視人道。重要的是，你必須知道組織成員能不能有效信守這些價值。我建議一項測驗：當市場、當你的產業、當你的客戶、當媒體都因為你抱持這些價值而懲罰你的時候，你仍然堅持不肯放棄的，是些什麼價值？只有這些價值，才是真正的核心價值。

How can we recognize what you call "core values" in an organization?

First of all, you don't need to have explicit core values. They don't have to be pretty, they can even be brutal, and they don't have to be humanitarian, although in most cases they are. The important thing is to know whether the values are believed in effectively. I recommend a test: What values would you continue to hold even if the market, your industry, your customers, and the media penalized you for holding them? Only such values are truly core.

————

照你這麼說，如果客戶與產業不讓一家公司信守它的核心價值，這家公司就應該放棄這些客戶與產業，即使這些客戶與產業能為它帶來利益也在所不惜？

完全正確。這是我們在書中刻意強調的一個很有震撼性

的理念，我們這麼做，為的就是要讓讀者停下來說：
「我沒有看錯吧？」大多數人認為，為了策略性需求，
公司需要不斷調整它的價值。但偉大的公司採取的途徑
正好相反：如果為了執行一項策略，必須採取與公司核
心價值不符的行動，那麼無論這項策略能為公司帶來多
大利益，它都會放棄。

So a company should abandon clients and industries that prevent it from being faithful to its core values, even if those clients and industries are profitable?

Exactly. It's a strong idea that we stated on purpose to make the reader stop and say: "Did I read this right?" Most people think that you need to adjust your values to your strategic needs. Great companies go the other way around: they discard any strategy—no matter how profitable it might be—if it would require actions that are inconsistent with the company's core values.

——

你沒有談創意領導人？

創意得靠公司。如果我們討論索尼或 3M，我會明確告
訴你，創意是這兩家公司核心價值的一部分，而且這
不是他們從書本上看來的，創意流在他們的血液中，

寫在他們的歷史裡。不過有些公司，例如諾斯壯百貨
（Nordstrom）等，並不重視創意。但不重創意並不表示
這些公司沒有非常強的價值。「顧客親密度」（customer
intimacy）也是一個商業用詞，它是一種策略概念，不是
核心價值。重點是，這世上沒有所謂「對」的價值，只
有可靠的價值。

What about being an innovation leader?

Innovation depends on the company. If we're talking about Sony
or 3M, I'll tell you clearly that innovation is a part of their core
values, that they didn't read it in a book. It's something that's
in their blood and written in their history. But there are other
companies that don't value innovation, such as Nordstrom, but
this doesn't prevent these companies from having very strong
values. Customer intimacy is also a buzzword. It's a strategy
concept, not a core value. The point is, there are no "right" core
values. The key question is not what are the "right" values, but
rather, what are the authentic values.

———

常有人將利潤最大化（maximizing profits）視為資本主義的
本質，這也是一家公司的宗旨嗎？

說這話的人錯了。我既不是社會主義者,也不是批判利潤最大化的第一人。我非常敬仰的管理大師彼得‧杜拉克(Peter Drucker),早在 1954 年就已經撰文指出,企業所以存在,為的不是利潤最大化。利潤最大化是一種反社會、不道德的東西。公司之所以存在,為的不是將股東或企業老闆的利益擴大到極限。公司當然必須顧慮獲利與賺錢的問題,利潤是公司的血液、氧氣與營養,但公司必須有更深一層的宗旨。不過,想找出一個宗旨,有時非常困難,許多公司於是用「利潤最大化」這項表述,來取代尋找宗旨這層意識型態上的麻煩。

What about maximizing profits, which is often termed the essence of capitalism? Is that the purpose of a company?

Those who say this are wrong. I am neither a socialist nor the first to say it. In 1954, Peter Drucker, whom I most admire, wrote that maximizing profits is not a reason for business to exist. It is antisocial and immoral. A company doesn't exist to maximize its shareholders' or business owners' profits. Of course it has to worry about earning profits and being profitable. Profits are the blood, oxygen, and nourishment of the company, but the company has to have a deeper purpose. The expression "maximizing profits" is an ideological substitute for the need to find a purpose, something that is sometimes very difficult.

但沒有一位有魅力的領導人，怎麼可能建立一個有遠見的組織？許多年來，商學大師們向我們推銷的概念，一直與這正好相反。

這些商學大師們完全錯了。我可以給你一張名單，列出 20 家世界級的公司，它們都沒有魅力型領導人。且讓我問你一句話：這些大師口中這種魅力領導觀念，是從哪裡來的？依我看，那不過是 20 世紀管理學版本作祟罷了，就像前一世紀的人假借神的名義，解釋一切搞不懂的問題一樣。魅力領導是一項成功要素，但成功還得靠其他要素。我們必須拋開諸如「魅力型領導」（charismatic leadership）這類簡化、一體適用式的答覆。

How is it possible to build a visionary organization without a charismatic leader? Business gurus have been trying to sell us the opposite for years.

These gurus are completely wrong. I can give you a list of 20 world-class companies that don't have a charismatic leader. Let me ask you this: where have all these mentors taken this idea of charisma? For me, it is just a twentieth-century version, for the management field, of what an earlier century tried to do by evoking God's name to explain everything. Charismatic leadership is one of the success factors, but there are others. Simplistic, all-purpose answers have to be discarded, like the charismatic leadership one.

———

既然如此，想建立一個基業長青的企業，有沒有什麼基本規則？

首先，每一家公司必須有一套核心意識型態——這是建立公司願景的第一要件。核心意識型態不像化妝品，不能敷上即可，也不能從其他地方抄襲、複製而來。它必須透過發掘而來。核心意識型態包括核心價值與核心宗旨，儘管時代轉變，這些價值與宗旨永遠不變。

其次，公司必須有一種對未來的願景，也就是說，必須為今後 10 到 30 年訂定遠大的目標。這些目標不必百分百可行；即使只有五成到七成勝算也沒關係。不過，想用這些願景打動人心，公司必須將它們說得栩栩如生。在描繪這些願景的時候，圖像必須明確、必須動人，必須激起人們的熱情與信念。

最後，如何將願景與實踐搭配，也非常重要。營造一家公司需要 1% 的願景，與 99% 的實踐配合——沒有這 1% 願景，其他一切都不重要。願景為公司帶來運作的天

地，但有了實踐的配合，不必看文件或廣告，也不必聽
「高層」演說，你也能了解這是一家什麼公司、它的目
標是什麼。你只要看它的營運與行動就可以了。

So, are there some basic rules for becoming a Built to Last enterprise?

First, every company needs to have a core ideology—this is the first component of vision. It is not applied like a cosmetic or copied from somewhere else. It has to be discovered. Having a core ideology means having both core values and core purposes. These are something that remain throughout time.

Then, it is necessary to have a vision of the future, which means defining ambitious goals for the next 10 to 30 years. These goals don't have to be 100 percent manageable; maybe they just have a 50 to 70 percent probability of success. But they'll have impact only if they're described in a lively way, if the images picturing them are clear and motivating enough, if there are passion and conviction.

Finally, aligning vision and implementation is essential. Building a company requires 1 percent vision (without it, nothing matters) and 99 percent alignment of that vision with implementation. Vision provides the context, but alignment allows anyone to understand what the company is about and where it is headed, without having to read papers or brochures, or listening to "top management" speeches. All you have to do is look at the company's operations and actions.

———

《從 A 到 A⁺》是續篇嗎？

事實上，我認為它應該是前篇。理想的情況是，讀者應該先讀《從 A 到 A⁺》，之後再讀《基業長青》。我們當時不知道，不過這兩本書加起來，可以完整訴說一家新公司怎麼成為好公司，然後成為偉大的公司，之後再成為歷久彌新、有遠見的公司。

Is *Good to Great* a sequel?

Actually, I see it as more of a prequel. Ideally, you should read *Good to Great* before you read *Built to Last*. We didn't know it at the time, but the two books, combined, tell the story about how a new company can become a good company, then a great one, then one that is an enduring, visionary company.

———

你認為「好」是「偉大」的敵人。為什麼？

「好」是「偉大」的敵人。社會之所以沒有偉大的學

校，是因為我們已經有很好的學校；我們之所以沒有偉大的政府，是因為我們已經有很好的政府；我們之所以沒有許多偉大的公司，也是因為我們已經有一大堆很好的公司。最重要的是，許多人不能享受偉大的人生，是因為他們覺得現在的生活已經不錯，無意繼續進取。這話說來可悲，卻是實情。想建立一家偉大的公司，就必須堅持極度嚴苛的標準。不能因行為與表現已經還不錯就躊躇滿志，不思進取。我們在這本新書中，提出幾項重大發現，其中一項就是：公司與經理人「不再」做什麼，永遠比他們做什麼更重要。

But you see "good" as the enemy of great. Why?

Good is the enemy of great. Society doesn't have great schools because we have good schools; we don't have great government because we have good government; and we don't have that many great companies because too many companies are simply good. Ultimately, it's sad but true that many people don't have great lives because they're willing to settle for good lives. To be a great company, you have to adhere to relentlessly stiff standards. You have to stop accepting good-enough behavior and performance. One of the major findings in the new book is that what companies and managers stop doing is infinitely more important than what's on their "to do" list.

―――

「偉大基因」深植於所有公司、經理人與員工嗎？

任何一家公司——我的意思是任何一個組織——都可以成就偉大的功業。事實上，這也是我個人在過去 10 年來的強烈感受。我覺得，我們已經確切知道，好的公司怎麼做才能成為偉大的公司。不過，好公司的員工必須從領導人以降全力投入，而且必須不懈地投入，才能達到這個目標。

Is the "greatness gene" embedded in all companies, managers, and employees?

Any company—and I mean any organization—can become a great one. That is truly one of my own epiphanies in the last decade. I feel that we learned exactly how good companies become great ones. But the people inside a good company, starting with the leaders, have to commit to this—and stay committed.

―――

第 4 章

真誠領導

《如何讓人願意被你領導》作者

羅伯・高菲、賈瑞斯・瓊斯

杜克企業教育學院執行總監

與談人 | **莉茲・梅隆**

21 世紀之初，領導人在道德標準上發生了幾項重大疏失，引起領導圈強烈反應。領導圈說，許多企業領導人道德淪喪，已經沒有行為標準，不再具備領導的真正要件，他們已經喪失自我。

在安隆（Enron）事件爆發，導致主張強化企業監控的「沙賓法案」（Sarbanes-Oxley Act）後，時任醫療產品公司美敦力執行長的喬治，眼見企業治理狀況頻出，於是主張培養一種更有道德的新型領導人。他說：

> 我們發現，今天企業欠缺的，是一種能為建立長期真誠組織而貢獻心力的領導人……我們需要真誠、可靠的領導人，這種領導人極度正直、願意全力投入，營造基業長青的組織。我們需要具有深度使命感，堅持企業核心價值的領導人。我們需要有勇氣、敢於面對所有員工、股東、客戶需求的領導人，我們需要重視社會服務的領導人。
>
> We realize that the missing ingredients in corporations are leaders committed to building authentic organizations for the long-term. . . . We need authentic leaders, people of the highest integrity, committed to building enduring organizations. We

need leaders who have a deep sense of purpose and are true to their core values. We need leaders who have the courage to build their companies to meet the needs of all their stakeholders, and who recognize the importance of their service to society.

———

　　喬治認為，所謂「真誠、可靠的領導人」（authentic leaders）具有幾種特質，特別是透過經驗而形成的價值觀，以及絕對的正直。這種價值觀能為領導人提供一種道德方針，而絕對正直能夠建立信任，為追隨者與員工灌輸一種使命感。這些概念在喬治 2004 年出版的《真誠領導：再揭創建不朽價值之祕》（*Authentic Leadership: Rediscovering the Secrets to Creating Lasting Value*）一書中有詳細論述，他在書中將這些理念進一步整理，找出 5 種特質，還為每種特質找出一種相關的附帶性發展特質。

　　喬治認為，真誠領導人了解他們的使命，同時也因為熱愛工作，對他們的使命充滿熱情。真誠領導人擁有明確的價值觀，而正直是其中不可欠缺的一部分，並且會在不同情勢下，身體力行。真誠領導人懂得以真心領導，能夠用熱情對待部屬，鼓勵員工成就偉業。真誠領導人會建立共同使命，營造一種息息相關的意識，從而與員工建立長

久關係，激發員工的忠誠與信任。真誠領導人極度自律，能夠有效面對工作壓力與緊張，常保身心健康。

喬治也探討其他一些有關領導的議題，包括如何營造真誠領導的公司、績效評估、治理與道德、創新、接班，以及不同層級的領導等。

▶ 真誠、正直、做自己

在對年紀介於 23 到 93 歲之間的 125 位領導人進行訪談之後，喬治與彼得・席姆斯（Peter Sims）於 2007 年出版了《領導的真誠修練：傑出領導者的 13 道生命練習題》（*True North: Discover Your Authentic Leadership*）一書。這 125 位領導人之所以為他們的訪談對象，主要是因為他們都以真誠、有效領導聞名。在當時，這是一次規模最大的有深度的領導學研究，進行這項研究的目的，是了解這 125 位領導人如何養成他們的領導能力。但兩位作者很快就發現這項事實：「分析完 3,000 頁的訪問稿之後，我們的團隊訝於發現，這些領導人認為他們的成功，與一般認為的成功個性、特質、技巧或風格完全無關。」

　　喬治說，這些領導人認為他們的領導能力，來自他們的人生故事：「他們透過現實經驗，有意識、無意識地不斷自我測試，不斷整理他們的人生故事，以了解他們最核心的真實自我。在這個過程中，他們發現了領導宗旨，知道真誠讓他們更加有效。」在展開真誠領導之旅以前，領導人必須先了解自己的人生，因為人生故事可以為真誠領導提供情節與環境，讓領導人根據現實生活教訓，進行當前與日後的領導。

　　根據喬治的理論，真誠領導的第一項能力，就是自我認知。喬治指出，要發掘真正的自我，領導人必須進行富有挑戰性，有時還令人十分痛苦的反思自省。不過，有許多領導人一心一意只關心自己的職涯前途，忽略了這種必要的反省工夫。此外，也有許多領導人只知道以股價、身分地位、職銜、金錢與名譽等外在標準來評估成功，卻從不考慮這些標準對他們是不是真的有意義。在坦然面對人生、誠實檢驗之後，領導人會比較謙和、比較有人味，不會那麼難以親近、高不可攀。

　　在附加誘因與本質誘因之間取得平衡，是真誠領導的又一要件，而這項要件也與自我認知息息相關。所謂平衡附加誘因與本質誘因，就是用外在標準來評估成功的同

時，領導人還必須考慮根據人生體驗而取得的本質性成功評估標準。喬治說，這類本質性成功評估標準，包括「個人成長、協助他人成長、投入社會公益、為改善世界盡一份心力」等。在附加誘因與本質誘因之間取得平衡，是真誠領導的重要關鍵。

建立平衡的人生並不簡單。領導人必須將人生各個層面結合在一起：家庭與朋友、職場生活與同事、商業性與社會性生活。喬治說，領導人必須結合這種種人生層面，並在每一種層面，都能展現真實的自我。唯有這麼做，才能展現前後一致的穩定風格與行為，才能建立信任。

真誠領導人透過與他人的合作取勝，因此真誠領導的另一項基本要件，就是建立「非凡的支援團隊」，協助真誠領導人保持正軌。這些團隊像顧問一樣，會在不同的時機提供新觀點和明智的建議，為走錯方向的領導人指點迷津。如果有必要，他們還能提供貼心慰藉與無條件的支持。而支援團隊的成員來自各方，包括家人、密友與同事。

喬治強調，要建立這樣的支援網絡需要時間。這類網絡一開始只有一名成員，隨著時間堆疊，成員人數也不斷增加。支援網絡必須是一種雙向關係，團隊支援領導人，

領導人也必須支援團隊。此外，領導人只是透過本身的人生故事，發掘自己的真誠領導風格還不夠，還必須學習如何鼓勵身旁的同事及員工，讓他們也能成為自成一格的領導人。

▶ 最難的莫過於做自己

在真誠領導的世界，喬治並不孤單。歐洲學界雙人組高菲與瓊斯也大力鼓吹真誠領導。高菲是倫敦商學院組織行為學教授，瓊斯是英國廣播公司（BBC）人力資源與內部通訊部主任，也是寶麗金唱片公司（Polygram）資深副總，在學界身兼多項要職。高菲與瓊斯在一開始，都是以社會學者的身分踏入學界。兩人寫道：「我們想改變這個世界，之後卻發現想做到這點，我們得先進一步了解組織行為的撲朔迷離。我們進入大公司展開研究工作，在大公司的環境中，遇到各式各樣汲汲營營、非常優秀，也往往讓人感到困惑的同事。」

兩人早期的作品也涉及領導，但並非直接涉及。他們在 1998 年出版的暢銷書《為你的公司算算命》（*The Character of a Corporation*）主要聚焦於企業文化，儘管如

此，這本書確實對領導人造成衝擊。高菲與瓊斯以社交力
（sociability）與團結力（solidarity）兩個文化特性為探討主
軸，社交力以人際交往為導向，團結力談的主要是對共同
目標的付出。兩人以 2×2 的座標，找出 4 類組織文化：網
路式、傭兵式、片斷式與公社式。每一類文化對領導人的
影響，都略有不同。

　　高菲與瓊斯從這一點出發，注意焦點開始轉向領導。
兩人的研究完全以領導現實為根據，與英雄式領導那套東
西無涉。兩人在 2000 年發表並贏得麥肯錫獎（McKinsey
Award）的〈為什麼別人要受你領導？〉（Why Should
Anyone Be Led by You?）一文中指出：「在事業上，如果
沒有人追隨你，你什麼事也做不成。在這個凡事講究『授
權』的時代，想找追隨者並不容易。因此，主管最好知道
怎樣才能有效領導──當主管，就得想辦法與人打交道，讓
人願意投入公司的目標。」

　　高菲與瓊斯認為，讓人樂於追隨的領導人，具有 4 種
主要特質：他們選擇性地暴露自己的弱點，使自己看起來
更加和藹可親；在決策過程中非常依賴直覺；用「嚴厲的
同理心」（tough empathy）管理員工；會適時彰顯自己與追
隨者不同的地方。不過，根據高菲與瓊斯的看法，有效領

導人最重要的一項特質，或許是他們強調真誠。高菲與瓊斯說：

> 想知道怎樣才能培養領導人，有太多值得探討的面向。在這眾多面向中，像這樣難做到的實在很少。這 4 種領導特質是必須採取的第一步，將它們加在一起，就是告訴主管：要真誠。在為主管提供建議時，我們總是說：「要用技巧，更進一步做你自己。」最難照辦的建議，大概莫過於此了。

> Of all the facets of leadership that one might investigate, there are few so difficult as understanding what it takes to develop leaders. The four leadership qualities are a necessary first step. Taken together, they tell executives to be authentic. As we counsel the executives we coach: "Be yourselves, more, with skill." There can be no advice more difficult to follow than that.

——

▶ 並非人人都能當領導人

在之後幾年，高菲與瓊斯潛心研究這個議題，不斷修正他們對領導的觀點。高菲與瓊斯說，首先，領導有一些

每個人都需要知道的基本事實。其中一個事實是，領導是理性的，所以如果要有領導，就得有追隨者。而且像大多數關係一樣，這種關係也需要耕耘。領導也是非階級的，換句話說，有一個響亮的頭銜，並不能讓你成為領導人。也就是說，只要有人願意追隨你，無論你置身於組織哪個層級，就算沒有響亮的職銜，你一樣可以成為領導人。根據高菲與瓊斯的看法，領導還是一種視情況變化的技巧，真誠的領導人雖然忠於自己，但還是會視情況需求來改變自己的行為。

領導人需要為追隨者，提供 4 種特別重要的東西。首先是社群：追隨者需要一種歸屬感，領導人可以幫助追隨者與他人結合。其次，領導人必須為追隨者帶來真實感。員工希望領導他們的，是真實的人，不是那種仰之彌高、可望不可及的領導人。追隨者希望領導人能適度自我表白，讓他們敢於親近。第三，領導人要讓追隨者覺得他們的努力很重要，若少了這份努力，組織便不會有今天的成就。最後，追隨者希望領導人能鼓舞、激勵他們，推動他們精益求精，更上一層樓。

高菲與瓊斯說，在實際作業上，領導人往往不知道如何真誠，也不知道怎麼樣才是真誠領導。在這種狀況下，

有 4 件事最為重要。隨機應變式的領導與真誠領導之間，有一種必然存在的緊張，領導人必須對組織狀況瞭若指掌，才能面對這種緊張而因應自如。高菲與瓊斯說：「只有能做到這點，你才能像一個『真誠的變色龍』一樣，隨環境有效調適，而不失自我意識。只有能做到這點，你才能採取象徵性行動與決定性行為，開始轉變情勢，為你的追隨者創造一種令人為之鼓舞的另類現實。」

高菲與瓊斯說，自我認知是真誠領導不可或缺的一部分，想做到這點並不需要進行長年的心理分析。自我認知的要旨，在於結合群眾共襄盛舉，領導人需要知道應該將哪些自我、多少自我、將自己的哪些長處與短處、哪些心理特質向他人表白。想結合群眾，真誠領導人就必須對追隨者非常了解——了解他們的希望、恐懼、興趣與情緒狀態。但要做到這一點，領導人首先必須接近追隨者。而在接近追隨者的同時，領導人有時不得不向同樣的追隨者提出挑戰，或是哄他們，甚或訓斥他們。面對這類情境，領導人需要知道如何與追隨者保持一定距離。

最後，為了讓追隨者接受他們表達的願景，同時還能感受到他們的真誠，領導人必須選擇最能切合他們個性與領導風格的媒體與時機。此外，他們表達的訊息也必須

清楚，讓每個人都容易了解。有意思的是，高菲與瓊斯同時詳述了有關領導的幾個流行概念，還說這些概念都是迷思。其中一個迷思，就是「人人都可以當領導人」，這與許多領導學專家的論點相反。高菲與瓊斯不認為每個人都可以當領導人，他們說：「許多主管並不具備領導必備的自知或真誠」；此外，許多主管根本也不想當什麼領導人。

▶ 如何管理聰明人才？

在完成真誠領導模式的修飾之後，高菲與瓊斯開始投入另一個許多領導人面對的棘手議題：如何管理桀驁不馴的聰明人。組織裡最有創意、最有生產力的一些人，往往不想當領導人，但也不願追隨他人的領導。對那些根據體制必須領導他們、帶領他們完成組織目標的人來說，管理他們著實很令人頭痛。

高菲與瓊斯說，領導人採取的方法如果得當，一切問題都能迎刃而解。千萬別告訴這些聰明人應該怎麼做，不要用階級、要用專家的地位展示權威。給這些聰明人一些空間，讓他們放手做他們最擅長的事。但如果他們提出

問題，你應該抽時間想辦法回答，並向他們提問。再怎麼說，他們往往比老闆聰明。聰明人喜歡獲得表揚與獎勵，但表揚與獎勵的方式，要對他們的胃口，特別是他們喜歡獲得他們崇拜的人（如該領域技術專家）的獎勵。此外，領導人必須真誠，這是當然的事。

▶沒有頭銜的真正領導人

多年來，我們與高菲與瓊斯會面過許多次，地點多在倫敦的酒館。我們在這些酒館裡高談闊論（特別是瓊斯），趣味橫生，每每流連忘返。兩人早在 1970 年代，還都是社會學學生的時候已經相識。

你們都是學社會學的，怎麼改行鑽研起領導了？

瓊斯：我們一直對真人實事很感興趣。為了撰寫〈為什麼別人要受你領導？〉，我們跨領域訪問了許多人，其中包括一位醫院護士、一位辛巴威軍人、一位小學校長、一位駕遊艇環球旅行的男子，還有其他各行各業在公司擔任大大小小各種職位的各色人等。我們很欣賞美國知名廣播節目主持人兼作家斯塔茲・特克爾（Studs

Terkel）的做法：與人交往，你會學習很多。他們在哪個領域領導無關緊要；無論在哪裡領導，他們都是領導人。

這與傳統做法不一樣。在過去，領導往往與英雄、名人形影相隨，但我們在公司內部進行的研究，看到的卻是各式各樣、精於鼓舞他人的領導人。真正讓我們心嚮往之的，是能夠擄獲人心、讓人心悅臣服的領導人。讓羅伯與我特別感興趣的，是德國社會學家馬克斯・韋伯（Max Weber）早已提出的那種反官僚與魅力的領導。對商界而言，具有這類特質的領導人，雖不是一切成敗勝負的決定性關鍵，但我們認為這類領導人非常有價值。

Your roots lie in sociology. How did you come to be diverted by leadership?

Jones: We have always been interested in real people doing real jobs. For "Why Should Anyone Be Led by You?," we interviewed a cross section of people. They included a hospital nurse, a Zimbabwean soldier, a head teacher, a round-the-world yachtsman, and a variety of others, as well as many people in an array of corporate positions. We like the approach of the great American broadcaster and writer Studs Terkel. If you engage with people, you learn from them. It doesn't matter where people exercise leadership; they are still leaders.

And that's different from the conventional approach. Leadership has tended to be associated with the heroic and the famous, but our work with companies has exposed us to a variety of leaders who excel at inspiring people. That's what really fascinates us: leaders who succeed in capturing hearts, minds, and souls. Rob and I are fascinated by leadership that, reaching back to the ideas of Max Weber, is antibureaucratic and charismatic. To have leaders with these qualities is not everything in business, but we think that it is worth a lot.

——

但領導總得有一些具體而且立竿見影的參數，不能只談態度與人格，對吧？

高菲：這話沒錯。領導必須講究成果，非這樣不可。偉大的領導能發揮激勵之效，讓人成就不凡。但領導涉及的，不只是業績表現，也包括意義。這一點很重要，卻往往為人忽視。各階層領導人都能提升績效表現，但他們之所以能做到這一點，是因為他們讓表現變得有意義。

而且，對社會、對個人而言，這種對意義的追求，已經愈來愈重要。隨著改變的步伐不斷加速，人們對長久、對意義的需求也日趨殷切。我們愈來愈擔心這個世界有

一天會為角色扮演人所主宰。

瓊斯：在組織裡，對意義、對領導人提供的凝聚力的需求，正益趨明顯。看一下階層架構就知道了。過去的組織世界，有各種巧妙的階層架構，職涯旅途或多或少尚稱穩定，疆界之分也很明確。現在，這一切都變了。問題在於，大家現在發現，階級不僅是組織裡的結構性協調裝置而已，更重要的是，它們還是意義的來源。血管裡流著公司血液的組織人，現在必須面對一個含糊不清、模稜兩可的世界。在這樣的世界，過度認同一個組織所帶來的，不是職涯，而是問題。階層架構愈來愈不明確，意義逐漸消失，我們於是仰仗領導人為我們的組織帶來意義。

這個過程已經持續了好一陣子。但最近幾年，不斷爆發的企業醜聞，使它再次成為世人矚目的焦點。這些醜聞是領導人道德敗壞的後果，對我們經濟凝聚的共識造成的傷害很大。這種道德敗壞的副作用之一，就是許多主管開始抱持一種憤世嫉俗的態度。如果你在這些主管上班的時候問他們：「什麼讓你的生命有意義？」他們會把公司最近流行的一套冠冕堂皇的說詞唸給你聽。如果你在他們回到家後問他們，他們會向你坦誠，工作壓力太大，家庭生活又搞得一團混亂，他們已經不知道人生

意義究竟在哪裡。

But surely leadership needs some hard-and-fast parameters. It isn't just about attitude and personality.

Goffee: True, leadership is about results. It has to be. Great leadership has the potential to excite people to extraordinary levels of achievement. But it is not only about performance; it is also about meaning. This is an important point—and one that is often overlooked. Leaders at all levels make a difference when it comes to performance. They do so because they make performance meaningful.

And the quest for meaning is increasingly important to societies and individuals. As the pace of change increases, individuals are ever more motivated to search for constancy and meaning. We've become increasingly suspicious of a world dominated by the mere role player.

Jones: In organizations, the search for the meaning and cohesion that leaders provide is increasingly clear. Look at hierarchies. In the old world of organizations, there were ornate hierarchies, more or less stable careers, and clear boundaries. All this has changed. The trouble is that people now realize that hierarchies were not just structural coordinating devices in organizations. Rather, and much more significantly, they were sources of meaning. The organization man, with company blood coursing through his veins, now has to come to terms with a world of high ambiguity in which overidentification with one organization is a problem rather than a career. As hierarchies flatten, meaning disappears, so we look to leadership to instill our organizations with meaning.

This process has been under way for a while. But the corporate scandals of the last few years have brought it into the spotlight. They are a symptom of amoral leadership, and the damage that they have done to the ideology that makes our economic system cohere has been substantial. One side effect of this is that there is a lot of cynicism among executives. If you ask them while they are at work, "What gives your life meaning?," they mouth the latest corporate platitudes. If you ask them at home, they will admit to profound symptoms of meaninglessness as they struggle with work-related stress and dysfunctional family lives.

——

領導與意義，兩者有什麼關聯？

高菲：如果沒有一個說得很清楚的宗旨、使命，意義自然也模糊不清，領導的目的就在於把宗旨說清楚、講明白。我們每在進入組織進行研究時，總是發現這種對真

誠、對領導的追求正不斷加強。執行長告訴我們，他們
的組織現在最迫切需要的，不是似乎充斥在他們身邊的
那些技藝高超的角色扮演人，而是更多領導人。位居組
織中下層的人士在與我們談話時，不是要求組織有更讓
人鼓舞的領導，就是渴望提升他們自己的領導技巧。真
誠領導，已經成為最珍貴的組織與個人資產。

瓊斯：我們在問組織員工，他們最希望建立哪一套能力
時，也發現這個事實。他們的答案都一樣：幫我們成為
更有效的領導人。他們眼見有效領導能為他們的人生、
能為組織表現帶來重大差異。當我們問執行長，他們面
對最大的問題是什麼時，結果也一樣。他們每個人都
說：我們組織在每一階層，都需要更多領導人。

What's the link between leadership and meaning?

Goffee: If there isn't a clearly articulated purpose, meaning is
elusive. Leadership provides that articulation. This search for
authenticity and leadership is reinforced whenever we work
inside organizations. CEOs tell us that their most pressing need
is for more leaders in their organizations—not the consummate
role players who seem to surround them. And among those who
are lower down in the organization, there is either a plea for
more inspiring leadership or, just as common, a fierce desire to

develop leadership skills for themselves. Authentic leadership has become the most prized organizational and individual asset.

Jones: That's what we find when we ask people which set of competences they would most like to develop. They all come up with the same answer: help us to become more effective leaders. They have seen that leadership makes a big difference to their lives and the performance of their organizations. The same is true when we ask CEOs what is the biggest problem they face. They unerringly reply: our organizations need more leaders at every level.

———

為什麼這麼欠缺領導人？

瓊斯：依照我們的看法，這有兩個原因。首先，組織或許渴望有更多領導人，但組織自我結構的方式讓領導無法生存。有太多組織的結構，在根本上就是領導毀滅機制。太多組織要不鼓勵因襲盲從，要不鼓勵角色扮演，而這兩者都不能培養有效的領導人。

第二個原因是，我們對領導認識不清。學界對領導已經做了那麼多研究，奇妙的是，我們對領導竟仍然那麼一無所知。我們這麼說，並不是對學界同儕的批判，但我

們對他們使用的研究方法，以及他們引用的基本假定感到懷疑。

高菲：看一看主流領導學術著作，你會發現它強調的是領導人的特質。這裡面有很強的心理偏見。他們認為，領導特質是個人與生俱來的天賦。他們的主要假定是，領導是一種我們「對」他人做的事。但依照我們的看法，領導應該是一種我們「與」他人一起做的事。無論在任何時候，我們都必須將領導視為一種領導人與被領導人之間的關係。

透過這種研究方式與假定寫出的領導書籍，難免千篇一律想為領導找一個妙方。這些書列了一大堆領導能力與領導特質。讀這些書的人，最後一定失望。讀前奇異執行長傑克・威爾許（Jack Welch）的領導書，就算讀得再熟，也不會讓你成為像威爾許這樣的領導人。

Why are leaders in short supply?

Jones: There are two reasons, we think. First, organizations may desire leaders, but they structure themselves in ways that kill leadership. Far too many of them are machines for the destruction of leadership. They encourage either conformists or role players. Neither makes for effective leaders.

The second reason is that our understanding of leadership is blinkered. For all the research into leadership, it is surprising how little we know about it. We're not criticizing our academic colleagues when we say that, but we are questioning the methods they have used and the fundamental assumptions upon which much of the research has rested.

Goffee: Look at the main leadership literature and you will see that it focuses on the characteristics of leaders. There is a strong psychological bias. It sees leadership qualities as being inherent in the individual. The underlying assumption is that leadership is something that we do to other people. But in our view, leadership should be seen as something that we do with other people. Leadership must always be viewed as a relationship between the leader and the led.

A corollary of this is that books on leadership persistently try to find a recipe for leadership. There are long lists of leadership competences and characteristics. Anyone who reads these books is bound to be disappointed. Reading about Jack Welch isn't going to make you into Jack Welch.

――――

照這麼說，這世上並沒有普世性的領導特性？

瓊斯：我們認為沒有。對一位領導人有效的特性，對另一位領導人未必有效。你如果想成為一位領導人，就得先發掘自己，知道自己一旦置身領導環境之中，能夠拿出什麼本領。

So there are no universal leadership characteristics?

Jones: We don't think so. What works for one leader will not work for another. If you want to become a leader, you need to discover what it is about yourself that you can mobilize in a leadership context.

————

你的意思是，想領導就得全面了解自我？

瓊斯：這確實是當代許多領導書籍的建議。毫無疑問，以 EQ 為例，能夠全面掌控 EQ 當然非常有用，但問題是，與我們談過話或與我們共事過的那些領導人，沒有一個人能全面了解自我。人生與領導並不是這樣的。

高菲：我們結識的這些領導人，有強大的使命感，以及足以了解本身領導潛能的自我認知。他們不能全面掌握，但他們知道得夠多。

瓊斯：這麼說或許有一點太功利，但事實上，這是根據 3 項領導原理而做出的結論。第一項原理是，領導隨情境變化而變化。情境永遠影響對領導人的需求，如 911 事件爆發時的時任紐約市長魯迪・朱利安尼（Rudy Giuliani）或邱吉爾，都是這類例子。在組織生活中，削減成本、轉虧為盈的強勢經理人，在組織需要擴建時，往往不能有效領導。

我們的第二項觀察心得是，領導與階級無關。攀升到組織高位，並不能讓你成為領導人。階級本身既不是領導的必要條件，也不是領導的充分條件。

高菲：想在大型、政治性往往非常濃厚的組織攀升高位，需要具備的特質，與領導並沒有明顯的關係。能在這類組織攀登高位的人，靠的往往不是真正的領導特質，而是各式各樣的原因，其中包括政治機智、個人抱負、服務年資，甚至裙帶關係等。

Do you mean that in order to lead, you need complete self-knowledge?

Jones: That's what a lot of the contemporary writing about leadership suggests. But, while it is undoubtedly very useful to have a great deal of emotional intelligence, for example, none of the leaders we have talked to or worked with have had full self-knowledge. Life and leadership aren't like that.

Goffee: What they do have is an overarching sense of purpose, together with sufficient self-knowledge to recognize their potential leadership assets. They don't know it all, but they know enough.

Jones: That might sound a bit too pragmatic, but it is actually based on recognizing three fundamental axioms about leadership. The first of these is that leadership is situational. What is required of the leader will always be influenced by the situation. Think of Rudy Giuliani in the wake of September 11 or of Winston Churchill. In organizational life, hard-edged, cost-cutting turnaround managers are often unable to offer leadership when there is a need to build.

Our second observation is that leadership is nonhierarchical. Reaching the top of an organization does not make you a leader. Hierarchy alone is neither a necessary nor a sufficient condition for the exercise of leadership.

Goffee: You could argue that the qualities that take you to the top of large-scale and often highly political organizations are not obviously the ones associated with leadership. People who make it to the top do so for a whole variety of reasons—including political acumen, personal ambition, time serving, and even nepotism—rather than real leadership quality.

——

所以，領導並不是少數天縱英明之士的專利。

高菲：不是。偉大的組織在所有階層都有領導人。無論是醫院、慈善團體或商業企業，成功的組織總是想方設法，全面營建領導能力，讓組成分子有機會領導。

瓊斯：我們的領導觀的第三項原理是，領導是理性的。簡單說，沒有追隨者，你當不成領導人。領導是一種由雙方積極建立的關係，這種關係網絡很脆弱，需要不斷再造、創新。

所謂理性關係，不表示一切事情必須永保和諧。事情不可能永保和諧，理性關係也可能出現緊張情勢，但那是因為有效領導人知道如何激勵部屬、讓部屬成就不凡所致。

So leadership is not the sole preserve of the chosen few.

Goffee: No. Great organizations have leaders at all levels. Successful organizations—whether they be hospitals, charities, or commercial enterprises—seek to build leadership capability widely and to give people the opportunity to exercise it.

Jones: The third pillar of our view of leadership is that leadership is relational. Put simply, you cannot be a leader without followers. Leadership is a relationship that is built actively by both parties. This web of relationships is fragile and requires constant re-creation.

This doesn't mean that everything is always harmonious. It isn't. There may be an edge in a relationship, but that's because effective leaders know how to excite their followers to become great performers.

——

就非常實用的層面而言，這對那些有志領導的人有什麼影響？他們需要知道什麼？需要做什麼？

瓊斯：答案很簡單，簡單得讓人難以置信。事實上，想成為一個更有效的領導人，你必須做你自己，而且運用技巧，更進一步做你自己。

首先，想當領導人，就必須做你自己。我們不妨觀察維京（Virgin）老闆理查‧布蘭森（Richard Branson）爵士，看一看他怎麼運用他的外貌——輕鬆隨意的穿著、長髮、鬍鬚——傳遞一種不拘形式、不受傳統束縛的訊息，讓這

種訊息成為布蘭森式領導與維京品牌的核心要件。追隨者希望領導他們的是人，不是一個角色、一個職位，或一個官僚。

我們列為研究對象的領導人，都非常懂得運用他們與眾不同的特質來吸引追隨者。我們研究理查‧布蘭森與眾不同的特質，得到一項訊息：特質必須是真跡，不是仿造品，而且還要讓其他人看得見。所以，我們談的不是任何人格差異，而是一種經常需要經過多年微調、修正才能有成，以巧妙手段真實展現、能夠鼓舞他人的真正差異。

高菲：自我認知與自我表白之間的關係，是了解有效領導的重要起點，而且愈來愈受到重視。但它不是領導的全部，領導人本身必須進入狀況。偉大的領導人能夠認清情勢，隨機應變。他們懂得運用既有資源，為組織帶來更多資源。套用管理學術語，他們能夠為組織添加價值。為了達到這個目的，領導人需要了解如何在本質與環境間做調適、如何在個人特性與組織一致性之間取得平衡。

有效領導人不僅是因應環境變化、採取行動而已。他們還懂得如何運用足夠的適應力來塑造環境。這是領導

力的技巧面要素，領導人必須知道在什麼時候、什麼地方適應環境。若不能具備這項能力，領導人不大可能生存，也很難與他人結合、建立成功的領導關係。想要有效領導，領導人必須讓自己的行為充分融入組織文化，以創造吸引力。不能做到這一點的領導人，只能在與追隨者隔離的情況下空轉，什麼事也做不了。

What are the implications—at a very practical level—for those who aspire to leadership? What do they need to know and do?

Jones: The answer is simple, deceptively simple, in fact: to become a more effective leader, you must be yourself, more, with skill.

First, to be a leader you must be yourself. Look at Sir Richard Branson, the Virgin boss, and the way he uses his physical appearance—casual dress, long hair, and a beard—to convey the informality and nonconformity that have become a central part of his leadership and, indeed, of the Virgin brand. Followers want to be led by a person, not a role holder, a position filler, or a bureaucrat.

The leaders we studied were very adept at deploying their differences in ways that attract followers. Richard Branson's differences signify a message; they are authentic, not falsely manufactured; and they are seen by others. We are talking, then, not of just any personal difference, but of an artful and authentic display, often fine-tuned over many years, of genuine differences that have the potential to excite others.

Goffee: The link between self-knowledge and self-disclosure is a central—and increasingly fashionable—starting point for understanding effective leadership. But it is not everything. Leaders must be themselves in context. Great leaders are able to read the context and respond accordingly. They tap into what exists and bring more to the party. In management jargon, they add value. This involves a subtle blend of authenticity and adaptation, of individuality and conformity.

The thing about effective leaders is that they do not simply react to context. They also shape it by conforming enough. This is the skill element. It involves knowing when and where to conform. Without this, leaders are unlikely to survive or make the connections they need in order to build successful relationships with others. To be effective, the leader needs to ensure that his or her behaviors mesh sufficiently with the organizational culture to create traction. Leaders who fail to mesh will simply spin their wheels in isolation from their followers.

———

你能不能解釋一下，你所謂「運用足夠的適應力」指的是什麼？

高菲：能夠成功改造組織的領導人，一定得挑戰組織既有規範，但在絕大多數情況下，領導人不能同時挑戰一切規範。在沒有摸清楚組織環境的情況下，他們不會貿然展開對抗。他們必須針對現行、既有的社會關係與網絡，進行有計劃地調適才能生存（特別是在改革行動初期，情況尤其如此）。想要改變組織現狀，領導人必須先以組織成員的身分，至少取得組織其他成員最低限的接納。而且，求得早期生存的規則，與取得長期成功的規則通常不一樣。

瓊斯：觀察企業世界，你會發現執行長不顧組織環境、肆意孤行的例子多得不勝枚舉。這些執行長有時也能取得一些短期利益，但就長期而言，領導人想推動持久性的改革，就得先了解並適應組織環境。在了解、適應組織環境以後，領導人才能以公信力為後盾來推動改革，改革成功的機率也因此較大。若疏忽這個步驟，後果可能災情慘重，像艾爾・鄧拉普（Al Dunlap）這類以鐵腕手段削支、裁員的執行長就是例證。他們有時也能取得短期成功，但就長期而言，多以大敗虧輸收場。

高菲：要點在於，誰能正確解讀組織，以及如何培養這種解讀技巧。很顯然，有些領導人主要由於擁有多年面對各種情勢的經驗，不假思索就能解讀情勢。他們已經學得一種智慧，在研判情勢、甚至在採取干預行動時，能夠比較不依賴概念模式。但有沒有一些能讓我們了解組織關係，能讓我們知道改革成功機率的一般性原則？我們認為有。我們的顧問經驗告訴我們，許多客戶已經找到能幫他們解讀情勢的模式。

我們根據一種將組織視為社群的觀點，擬出一套了解組織環境的辦法。我們主要以經典社會學理論為基礎建立這個模式，以社交力與團結力為文化關係的兩大主軸。社交力指的主要是友人之間的情感關係。這些友人一般能共享理念與價值，在平等條件上相互交往。就其核心本質而言，社交力代表一種為了社交而社交的價值關係。它通常以面對面接觸為開端，但可以經由其他溝通形式而維持，同時它還有高度互助的特性。社交力沒有真正的附帶條件。

相形之下，團結力是個人與群體之間以任務為重心的合作。它不必依賴親密的友人關係，甚至不必以人的相識為基礎，而且也不需要持續。它的產生，純粹出於一種共同利益概念。這種概念一旦出現，團結力能造成密集

的專注。

儘管這麼說，似乎有些抽象，社交力與團結力，其實充塞於我們身旁各處。在我們的家裡、在球隊、在社交俱樂部與社區，它們無處不在。有人認為，早期社會學者之所以注意到它們，正是因為它們這種無所不在的特性。事實上，我們每個人都與它們脫離不了關係，也都受它們影響。舉例說，隨便要一個人描述理想家庭，一般都會告訴你，家人要能夠相親相愛（社交力），遇到困難時要能夠同舟共濟（團結力），這就是理想的家庭。

Can you explain what you mean by conforming enough?

Goffee: Leaders who succeed in changing organizations challenge the norms—but rarely all of them, all at once. They do not seek out instant head-on confrontation without understanding the organizational context. Indeed, survival (particularly in the early days) requires measured adaptation to an ongoing, established set of social relationships and networks. To change things, the leader must first gain at least minimal acceptance as a member—and the rules for early survival are rarely the same as the rules for longer-term success.

Jones: If you look around the corporate world, there have been countless examples of CEOs who rode roughshod over organizational contexts. Sometimes they have reaped short-term gains. But, in the long term, sustainable change requires the leader to understand and tune in to the organizational context. Having done so, the leader can instigate change with credibility and with a greater chance of success. Ignore this, and the results can be disastrous. Think of Al Dunlap or the host of ruthless downsizers and asset strippers who conspicuously fail to deliver long-term change.

Goffee: The question is: Who can read organizations well, and how do they develop this skill? Clearly, some leaders are able to read situations intuitively largely as a result of many years' experience in different contexts. They develop a kind of wisdom that makes them less dependent on conceptual models to give them insight or even to guide their interventions. But are there universal principles that underlie organizational relationships and that might frame the possibilities for change? We think there are. Our consulting work suggests that many people find models that refine their context-reading skills.

We have developed a way of understanding organizational context that is based on a view of organizations as communities. In our model, which draws heavily on classic sociology, there are two key cultural relationships: sociability and solidarity. Sociability refers primarily to affective relations between individuals who are likely to see each other as friends. They tend to share ideas and values and to associate with each other on equal terms. At its heart, sociability represents a relationship that is valued for its own sake. It is usually initiated through face-to-face contact, although it may be maintained through other forms of communication, and it is characterized by high levels of mutual help. No real conditions are attached.

Solidarity, by contrast, describes task-focused cooperation between individuals and groups. It does not depend on close friendship or even personal acquaintance, nor does it need to be continuous. It arises only from a perception of shared interest—and, when this occurs, solidarity can produce intense focus.

Although this discussion may seem a little abstract, sociability and solidarity relationships are actually all around us—in our families, sports teams, social clubs, and communities. Arguably, this ubiquity is what drew the attention of the early sociologists in the first place. In effect, we all have an interest in—and are affected by—these relationships. Ask someone to describe his or her ideal family, for example, and typically that person will tell you it is one in which the members like and love one another (that's the sociability element) and that pulls together when times get tough (that's solidarity).

———

領導生活中有太多緊張與矛盾，領導人必須非常敏感才行。

瓊斯：沒錯，確實有很多緊張。領導人必須展現自己的長處，但也要暴露自己的弱點；要有個性，但同時要有足夠的適應力；要建立親密關係，但也要與部屬保持距離。管理這些緊張，是成功領導的核心要件。

There are a lot of tensions and paradoxes at work here. The leader needs to be incredibly sensitive.

Jones: Yes, there are a lot of tensions. Leaders must reveal strengths, but show weaknesses; be individuals, but conform enough; establish intimacy, but keep their distance. Managing these tensions lies at the heart of successful leadership.

———

想做到這些事太難了。不如模仿傑克‧威爾許還比較容易些，不是嗎？

高菲：模仿威爾許的問題在於，就算你真的可以模仿，對威爾許有效的做法對你未必有效。

我們的經驗告訴我們，即使能夠精通一、兩項我們談到的這些領域，仍不足以成就真正令人奮發、鼓舞的領導。偉大的領導人之所以能在適當時機、用適當風格進行領導，是因為他們能夠覺察情勢變化，而且懂得運用這些領域之間的交互作用。換句話說，每位領導人都是獨特的。

瓊斯：而且領導之難，也非比尋常。把領導說成一件簡單的事，一點意義也沒有。任何曾經擔任領導職的人都可以告訴你，領導很複雜、很費神，而且充滿個人風險。十分顯然，並非每個人都可以當領導人，而且許多非常有才華的人，打從心裡就對肩負領導責任毫無興趣。一個能讓他人奮發、受到鼓舞的領導人，需要充沛的活力、衝勁與意志力，假設每個人都能具備這些條件，未免太草率了。我們的看法是，每個人都有一套與眾不同的潛能，可以運用在領導上。所以，我們每個人都必須回答一個不容迴避的問題：我要不要當領導人？如果要，這個願望是不是夠強，強到讓我願意付出必要的辛勞，願意做出必要的犧牲？

高菲：而且，一旦你成為領導人以後，你必須問：為什麼別人要接受你的領導？為什麼我們要讓你領導？有效領導人必須在每天說每件事、做每件事的時候，都能答

覆這些問題。如果做不到這一點，領導的落實難逃宿命
的失誤，領導人才荒也將持續。

That's quite demanding. Wouldn't it be easier to imitate Jack Welch?

Goffee: The trouble is that even if that were possible, what works for Jack Welch won't work for you.

Our experience suggests that excellence in one or two of these areas we've talked about is insufficient for truly inspirational leadership. It is the interplay among these areas, guided by situation sensing, that enables great leaders to find the right style for the right moment. In other words, every leader is unique.

Jones: And leadership is uniquely difficult. There is no point pretending that leadership is straightforward. Anyone who has ever been in a leadership position will tell you that it is complicated, demanding, and full of personal risk. Clearly, not everyone can be a leader, and there are many very talented individuals who are simply not interested in shouldering the responsibility of leadership. To assume that everyone has the sheer energy, drive, and willpower required to offer inspirational leadership to others is facile. We argue that each individual has unique differences that potentially can be exploited in a leadership role. So, each of us has to address the blunt question: Do I want it? And if I do, do I want it enough to put in the work required and make the necessary sacrifices?

Goffee: And then if you take on a leadership role, you have to ask: Why should anyone be led by you? Why should we be led by you? Effective leaders must answer these questions every day in everything they say and do. Otherwise, the shortage of leaders will continue, as their practice of leadership will be fatally flawed.

――

▶ 領導人的想法

真誠是一種令人心安的特質。我們都能做自己,這當然不成問題,但要做到真正的真誠,卻難得出奇。誠懇與虛假、對與錯、真實與偽詐都涉及判斷,也涉及品味。杜克企業教育學院(Duke Corporate Education)執行總監,同時也是《領導人的內心世界:像領導人一樣思考的 5 種方式》(*Inside the Leader's Mind: Five Ways to Think Like a Leader*)作者莉茲・梅隆(Liz Mellon),對領導的真誠問題,有不同的省思。

為什麼特別注意領導人的心靈？

我在這一行做了二十幾年，花了許多時間講課或為領導人提供訓練。在這許多年間，有個問題一再反覆出現。這個問題追根究柢，就是真誠的問題。

我的公司或組織，對於卓越的領導有一套觀點，也有一份領導特質與能力的清單，我要怎麼做，才能遵循公司或組織的模式，而仍然身為真誠的領導人？我要怎麼做，才能一方面身為他們期待的企業領導人，另一方面仍在本質上保有自我？

Why focus on the leader's mind?

I had been in the business for a long time, and had spent such a lot of time in the classroom or in coaching sessions with leaders, and one question kept coming up again and again. It was really about authenticity.

How do I follow my company or organization's view of the excellent leader—its list of leadership attributes or competencies? How do I follow that and still be an authentic leader? How can I still be, in essence, me as well as the corporate leader they want me to become?

——

這問題對妳有什麼啟發？

它讓我知道，我們有疏失。我們雖然想盡辦法觀察、評估領導人的行為，但仍然疏忽了一項非常重要的東西。

這項東西就是領導人的思考。談到行為，領導人有各式各樣不勝枚舉的行為模式，想列出一張清單以便掌握，非常困難。領導人之所以掙扎，原因就在這裡。他們望著這張列得密密麻麻的清單，不知道自己究竟在哪裡。

但我們的行為，靠的是我們怎麼思考，怎麼觀察一個情勢，怎麼將我們的工作概念化，怎麼看待這個世界。所以說，領導人在清單上的位置，取決於他們如何思考自己，以及自己的工作。

What did that tell you?

It told me that we are missing something. In our drive to observe and measure how our leaders behave, we are missing something very important.

It's about how leaders think. When you get to behavior, there's such a huge range of different ways in which leaders can behave that when you try to capture it in a list, it is difficult. That's why leaders struggle. They look at the big list and wonder where they are on it.

But how we think, how we see a situation, how we conceptualize our job, how we believe the world works—that's what drives our behavior. So where leaders are is based on how they think about themselves and about their job.

——

妳談到 5 種像領導人一樣的思考方式。

是的，不過首先，我要聲明一件事。這 5 種方式沒有任何先後順序。領導人的思考不是「我要先做這個，再做那個」，它比較像是舞步。你知道舞步，但要選擇什麼音樂、怎麼跳，取決於你，取決於個人。所以說，行為方式各有不同。

You describe five ways of thinking like a leader.

Yes, but the first thing to say is that they don't go in any order. It's not, "Right, first I'll do this, then I'll do this." It's more like dance steps. You know the steps are there, but the music and the pace will depend on you, the individual. So, how you get there varies.

——

妳舉出的 5 種思考方式中，有一種叫做「沒有安全網」（no safety net），這是什麼意思？

這就是說，既然身為領導人，就得踏出去。面對生死存亡關頭，走到懸崖邊緣，每個人都指望領導人在沒有安全防護網的情況下，第一個踏出去。沒有安全網保護他，領導人必須先踏出去，必須想辦法保持平衡。他必須這麼勇敢才行。

我問那些接受我訪談的人，所謂「沒有安全網」是什麼意思，他們說那是一種思考方式：「那表示，我必須是第一個到那裡的人。踏出第一步的人是我，冒這個風險的人是我。採取這個行動的人是我。」

不過，當我要他們舉例說明，如何踏出這個第一步時，他們的回答就五花八門、無奇不有了。

One of the five ways of thinking is called *no safety net*. What does that mean?

That means that someone has to step out because he or she is the leader. Everyone expects the leader to be the first one to step out—over the precipice, over the abyss, with nothing below him. The leader is the one doing the balancing act. There is no safety net; he has to step out. He has to be that brave.

When I talk to the people I interviewed and ask what *no safety net* means, they describe it as a way of thinking: "It means that I have to be the first one out there. I'm taking the first step. I'm taking the risk. I'm the one who does that."

However, when I ask them for examples of how they do that, I get a huge range of responses about behaviors.

——

其他類型的思考方式呢?

有一種非常個人的思考方式,叫做「在不適中求舒適」(comfortable in discomfort)。這世上有太多模糊不清、複雜多變的事。假設你經營一家跨國公司,你的生意遍及許多國家,旗下有好幾千名員工,你得面對市場、政治、民族文化等各式各樣複雜的問題。

你不得不接受這些現實,想辦法面對這一切。你坦然面對這紛至沓來的重重困擾,勇於採取必要的決策,這就是「在不適中求舒適」。

這種思考的呈現方式各有不同。有時,領導人已經擁有一大堆資料,知道自己的決定錯不了:該不該前進中國,該不該撤出斯里蘭卡等,但知道時機還沒有到,所

以還是選擇按兵不動。也有時，領導人手邊一點資料也沒有，但仍舊下定決心採取行動。

What other types of thinking are there?

Another one that's at the very personal level is *comfortable in discomfort*. There's a whole lot of ambiguity and complexity out there in the world. Imagine that you are running a business that spans multiple countries, has thousands of employees, and faces all the complexity that goes with that: markets, politics, national cultures.

Somehow you have to live with that, find a way through it, and still have the courage to take the decisions that need to be taken, while being comfortable with the level of complexity that's coming at you.

That manifests itself in different ways. Sometimes the leader has tons of data to support her decision making: whether she should go into China or divest Sri Lanka or whatever the decision that she is considering may be, but she waits because she just knows that the timing isn't right. Or maybe there are no data at all, and the leader makes a decision to go in a certain direction.

———

對於必須應付這種複雜情勢的領導人而言，要他們不要把自己的壓力轉嫁給員工，一定非常困難。

完全正確。領導人必須日復一日，承受那種不確定的狀態，而且還得面帶笑容，以免身邊的人也必須像他們一樣，面對亂局而憂心忡忡。

我因此發現，領導人有一種我稱為「我就是企業」（I am the enterprise）的思考方式：必須非常仔細思考你向其他人表達了什麼訊息。採礦業集團力拓（Rio Tinto）前執行長湯姆‧艾班尼斯（Tom Albanese），就是一個很好的例子。在前後幾年之間，除了其他許多事情不計，他挺過一次敵意接管，買下一家規模幾乎與力拓一樣大的公司，還揮軍進入中國市場。他面對難以言喻的複雜與混亂，日復一日，從無間斷。

艾班尼斯告訴我一個非常有趣的小故事。在那段期間，某天開完一次極為惱人的會議，從會議室走出來的時候，他發現助理面帶愁容。於是，他對這位助理說：「我知道我必須面露微笑，但事實上，妳也必須面露微笑；因為妳若一幅愁眉苦臉，他們會心想，我一定對妳說了什麼壞消息，讓妳擔心。」

所以，就企業這個層級而言，管理一家公司的領導人，

要有一種「我就是企業」的思考方式，隨時注意自己與
自己的團隊帶給他人的訊息。

For leaders who are coping with that kind of complexity, it must be difficult not to convey the pressure that they have to deal with to their employees.

Exactly. The leaders have to live with that uncertainty, and with a smile on their face, so that they are not worrying everybody else around them with the level of ambiguity they are actually coping with, day in and day out.

That brings me to a way of thinking that I call *I am the enterprise:* thinking very carefully about the message you're sending out to others. A good illustration of this is Tom Albanese, the former CEO of Rio Tinto. In the space of a few years, among other things, he faced a hostile takeover, bought a company almost the same size as Rio Tinto, and entered the Chinese market. He faced a huge amount of complexity and ambiguity, and he had to be always on.

And he told me a really interesting anecdote. Coming out of one of the more fraught meetings during that period, he looked at his assistant and realized that she was looking a little bit worried. So he said to her, "I know I've got to be smiling, but actually you have to be, as well; otherwise they'll think I've told you something that's worrying you."

So at an enterprise level, and in running the organization, there is that way of thinking where the leader is always aware of the message that he and his team are sending out.

———

妳還提到一種稱為「由我當班守夜」（on my watch）的思考方式，又是什麼意思？

這也是一種組織層級、非常有趣的思考方式。取這個名字是因為，想像你駕著一艘遊艇、順著季風橫渡大西洋。船上其他人都已入夢，只有你必須睜大眼睛守夜。

「由我當班守夜」的意思是，在一段期間，你身肩大得出奇的責任感。你知道，在你之前另有一人值班守夜，你負責這一班，時間到了以後，你要把責任交給另一個人。所以你必須兼顧 3 個時段：過去、現在與未來。

經營一家企業的問題在於，今天的事就能令你忙到暈頭轉向，迫使你全力投入，讓你無暇顧及明天。

資誠聯合會計師事務所（PricewaterhouseCoopers, PwC）全球董事長丹尼斯‧納利（Dennis Nally），對我說了一個有趣的故事。他說，當他年紀較輕、還在為建立聲望打拚

的時候，每天只為當天的表現發愁。直到獲得晉升，先後出任資誠美國與資誠全球事務總裁以後，他才發現自己的職責所在，其實是今天做了一些什麼或許不能在今天創造任何價值，但能奠下基礎、使其他人能在日後功成名就的事。這種謙卑的態度，真的讓人很感動。

What about the way of thinking that you style *on my watch*?

This occurs at the organization level. It is a really interesting way of thinking. The image I have in my mind here is that you are crossing the Atlantic on a yacht, following the trade winds, and you are the one who has got to stay awake for the night. Everyone else on the boat is asleep.

On my watch means that, for a period of time, you have that incredibly enhanced sense of responsibility. You know that somebody took the watch before you, that you're on for this watch, and that you're then going to hand the responsibility over to somebody else. So you have to be able to take care of three time zones: the past, the present, and the future.

The problem with running a business is that today is so busy, so hectic, and so preoccupying that there is a danger that you will spend all your time thinking just about today. That won't help you create a different tomorrow.

Dennis Nally, the global chairman of PricewaterhouseCoopers (PwC), told me an interesting story. He said that when he was younger, he used to just worry about his results today, because that was when he was making his reputation. When he was promoted to CEO of PwC in the United States, and then to global chairman, he realized that his job was to worry about the things that he is doing today that may not be creating any value today, but that lay the foundations for somebody else to be famous in the future. That humility was really very interesting.

——

領導人有時不僅必須放下手邊的工作，為未來進行籌劃，還得了解、尊重過去，知道哪些應該保留、繼續，哪些應該放棄，以及在什麼時候放棄。

完全正確。我認為，有些領導人之所以不善於推動改革，原因之一就是他們在上任之初，已經擬妥一個 1 百天行動計劃。他們在沙盤上畫一條線，口中唸唸有詞：「要拋棄過去，過去的事已經是歷史；現在是現在，我們要動起來。」

問題是，領導人說的這幾句話，對象是成千甚至上萬的員工。而公司能有今天這番基業，靠的正是這許多員

工多年來投入的心血。當領導人向他們展示自己對未來
新世界的願景時，這些員工很可能心裡會想：「且慢且
慢，你說得天花亂墜，未必跟我有什麼關係。你的未來
會不會算我一份，我哪知道？」

所以，對過去的尊重非常重要；領導人需要把過去的故
事融入未來。但同時你得記住，如果過去是一個拖累你
的包袱，讓你無法邁向未來，你也必須輕柔地將過去留
在過去，但手段必須輕柔。

**Not only do leaders have to stop doing what they're doing today
in order to gaze into the future, but they also have to understand
and respect the past, and to know which bits to keep and honor
and which bits to discard and when.**

Exactly. I think one of the reasons that some leaders are very
poor at leading change is that they arrive with their 100-day plan
and draw a line in the sand, and it is often along the lines of, "Let's
discard the past; that was then, this is now; off we go."

But you're addressing thousands, maybe hundreds of thousands,
of people who've invested years of their life in creating where you
are today. When the leader shows them her vision of the brave
new world, they are likely to say, "Well, wait a moment. I'm
not sure I want a part of that. I'm not sure you see me as part of

that."

So reverence for the past is really important; you need to integrate that story into a future. Yet you have to remember that if the past is going to stop you from moving to the future, you have to gently leave it behind—but to do so gently.

———

尊重過去，就能做到真誠嗎？

是的，真誠絕對與尊重過去密不可分。我稱真誠為「硬核心」（solid core），像其他許多思考方式一樣，「硬核心」也有許多面向。這是一個喧囂騷擾、充滿資訊科技，而且無處不在變化的世界。放眼全球，你看到日本發生大海嘯，你看到巨型公司垮台，也看到各種始料未及的事件。處於這種通信大漩渦中，你得想辦法進入自己的內心世界，告訴自己：「這個『硬核心』是我的指針，有了它，我才知道需要朝哪個方向前進。」

這一切講究的，就在於真誠。我認為真誠領導的重心，就在於這種硬核心，就在於擁有這種內心的沉穩。我不知道它從哪裡來，而且事實上，執行長們也都不知道它從哪裡來。有些執行長說：「我在這一行幹了 25 年、30 年，靠的就是這些經驗。」

丹尼斯‧納利說:「事實上,早從 15 歲起,我一直就是
這樣思考的。我一直有這種內在沉穩的意識。」非洲最
大銀行標準銀行集團(Standard Bank Group)副執行長希
姆‧夏巴拉拉(Sim Tshabalala)說,對他而言,所謂硬核
心就是他的價值,其中包括利用商業平台打造社會。夏
巴拉拉就是透過這種思考,先在南非建立據點,然後將
版圖擴及整個非洲。

所以,這是一種複雜的概念,但真誠領導人都有這種概
念。他們對生活與工作的態度及做法,都以這種真誠意
識為核心。

And that brings us to authenticity?

Yes. Authenticity is absolutely integral to this. I call it the *solid
core*, and like many of the other thinking styles, it is multifaceted.
This is a tumultuous world, full of information and technology
and change. Look across the globe; there are events like the
tsunami in Japan and huge companies going bust that you
would never expect. In this maelstrom of communication,
somehow you have to go inside yourself and say, "This *solid core*
is my compass, and this will give me a sense of where I need to
go."

That is about being authentic. That's where I think the heart of authenticity lies. It is in having this core of inner certainty. I don't know where it comes from, and, actually, neither do the CEOs. Some of them say, "I've been in this industry for 25 years, 30 years, so it's all that experience."

Dennis Nally said, "Actually, I've been this way since I was 15. I've always had this inner sense." Sim Tshabalala, the group deputy chief executive officer of Standard Bank Group, said that for him, it's about his values, and one of his values involves using the platform of business to build society, in this case first South Africa and then Africa more broadly.

So it's a complex concept, but they all have it. They all have this core of authenticity that surrounds the way they approach life and their work.

——

這麼說，妳在擬出這套思考方式以後，與各式各樣領導人討論它們，發現它們似乎都能與這些領導人共鳴？

確實是這樣。我們觀察領導人的行動，歸納出這一套原則，現在我們用這套原則反推回去。所以說，這套思考方式是以實例、以領導人的實際作為為基礎。

這 5 種思考方式，很能引起共鳴。每每我話還沒講完，

與我一起討論的領導人已經開口說：「哦，妳說得沒錯。」然後告訴我一個相關故事。

So, having formulated this set of ways of thinking, you then took them out and discussed them with a range of leaders, and they seemed to resonate with those leaders?

They did. We have observed leaders in action and derived these principles, and now we are applying them back. So it's really grounded in practice, in what leaders really do.

These five ways of thinking really resonated. I can barely get through them before people say, "Oh, yes," and tell us a story that relates to one of them.

——

這麼說，這世上有很多像這樣思考的領導人。我們不知道他們怎麼能、為什麼會這樣思考，也不知道他們什麼時候開始像這樣思考。但我們其他想當領導人的人又該怎麼辦？我們也能學習這種思考方式嗎？

這個問題又得回歸到真誠上。有些領導人會說，他們一直就是這樣思考的。不過，有些領導人說他們是學來的。「在不適中求舒適」就是一個很好的例子，標準銀行集團前執行長賈可・馬利（Jacko Maree）曾對我說，他

用的就是「在不適中求舒適」的思考方式，但他在 20 年前並不這麼思考。這是他隨著年齡漸長，逐漸成熟而學來的。

所以說，好消息是，這類思考方式可以透過學習而得。對世界各地有志領導的人而言，這些都是可以擁抱、可以學習的思考方式。

So there are leaders who think like this. We don't know how or why, or when they began thinking like this. But what about the rest of us who want to be leaders? Can we learn these ways of thinking?

That's where it comes back to authenticity. Some leaders would say that they have always thought like that. But others have said that they learned. Comfortable in Discomfort is a good example. Jacko Maree, the former group CEO of Standard Bank, said to me that he was Comfortable in Discomfort, but he had not used to be 20 years ago. He learned this one; it came with maturity.

So the good news is that these ways of thinking can be learned. For aspiring leaders everywhere, these ways of thinking can be embraced and learned.

第 5 章

魅力領導與
黑暗面

哈佛商學院教授
拉克希‧庫拉納
達特茅斯塔克商學院教授
與談人│**席尼‧芬克斯坦**

「魅力」（charisma）本是一個希臘字，意即「贈禮」
（gift）。在新約聖經中，「charisms」是聖靈（holy spirit）
贈予的禮物，它們包括智慧、知識與信仰，也包括行奇
蹟、「說方言」（speak in tongues）的能力，以及用來組織
與建造教會的能力。

德國社會學家、哲學家與政治經濟學家韋伯，將魅力
觀念視為權威與合法性的一種根源。他用「魅力」來描述
一個情勢，在這個情勢中，權威的來源不是規則或地位，
而是來自一種「對特定、極神聖使命與英雄主義的獻身，
或是一個人的模範性格、言行舉止，或他宣示或下達的命
令」。

根據韋伯的看法，魅力隨危機而至。人在陷於困境
時，就會盼望魅力型領導人出現，用令他們著迷的使命
感，用熱情和願景帶領他們走向安全。社會學與政治學者
已經投入多年工夫，鑽研魅力的性質與特性。魅力型領導
人的魅力特質，會隨領導人不同而互異，其中包括宏偉的
遠見與意識型態、英雄壯舉，以及鼓舞群眾的能力等。部
分學者認為，魅力型領導是一種關係性概念，取決於追隨
者的看法。

▶ 魅力四射的時代

　　然而長久以來，魅力始終與拿破崙、邱吉爾和甘地這類軍事及政治領導人形影不離。直到 1980 年代，才真正開始在管理與組織理論的研究上綻放光芒。哈佛商學院教授拉克希・庫拉納（Rakesh Khurana）指出：「在過去的商界，魅力不像今天這麼重要。在二次大戰結束後 30 年間……執行長一般都是那種在組織裡一步步升遷的組織人。」

　　庫拉納說，在李・艾科卡（Lee Iaccoca）於 1979 年出任克萊斯勒汽車公司（Chrysler）執行長之後，這種情況變了：「艾科卡以一種過去商界領導人無人能及的方式，鼓舞部屬、屢創佳績。他使克萊斯勒轉虧為盈，也因此成為美國家喻戶曉的英雄，開啟了一個以英雄式執行長掛帥的新紀元。」

　　克雷蒙・麥肯納學院（Claremont McKenna College）教授傑伊・康格（Jay Conger），是首先以組織環境中領導魅力為題，提出研究架構的一位領導理論學者。他提出的研究架構，第一步就是要「摘除魅力的神祕光環」。根據康格的理論，魅力不是什麼神奇的超能力，而是一種行為過

程。將魅力視為一種因受他人影響而具備的特質之後，就可以針對這類具有魅力的行為進行分析了。康格認為，魅力型領導有幾項行為要素，他強調這些要素都相互關聯，形成一個「要素星座」。

這些要素在一個分為三階段的魅力影響過程中運作。首先，執行這項過程的領導人，必須評估現有情勢，決定需要使用哪些資源，並評估進行途中可能遇上的障礙。領導人同時必須觀察組織成員，評估他們的需求與他們的滿意度。在這個階段，魅力型領導人必須對他們的環境具有高敏感度，包括社會環境與實體環境。在具備這種敏感度以後，他們才能對環境中的限制、達成目標所需的資源、組織成員的心態做出正確評估。正確評估是關鍵要素，因為魅力型領導人往往必須冒險，也必須採取激進的行動，讓每個人都能追隨自己一起行動。

在這個階段，還有一項行為要件，就是在現有系統中找出瑕疵與失誤的能力。魅力型領導人「積極尋找現狀中既有或可能出現的缺失」。由於能夠找出組織缺失，而且不肯容忍它們得過且過，這類領導人是組織的改革派，是改革的推動人。

在魅力影響力的第二個階段，領導人要訂定並表達目標。魅力型領導人之所以不同於其他領導人，既因為他們訂定的目標類型，也因為他們向他人表達這些目標的方式與眾不同。康格認為，魅力型領導人對未來具有遠見，這種遠見就本質而言，一般都有濃厚的策略性、都很理想化，與現狀有很大的差異，而且能體現組織成員的共同觀點。為了表達這種遠見，魅力型領導人必須是一個能讓人信服的人，就這點而言，「討喜」很重要。康格說，魅力型領導人必須，或至少「看起來」是「一位討喜、值得信任，而且知識豐富的人」。

最後，為了陳述、傳播目標，條理分明的表達能力是基本要件。魅力型領導人必須將現有與未來的情勢，以及自己的領導動機說清楚。這麼做的同時，還得小心翼翼，一方面強調自己的遠見有多好，另一方面也要強調現狀的負面因素。

在魅力影響力的最後階段，領導人要讓人知道組織應該怎麼做，才能達到他／她訂定、而且為組織全員共有的目標。為了做到這一點，魅力型領導人需要讓人相信自己「值得信任」。魅力型領導人能夠向追隨者表達自己對現狀的不滿，也能夠將自己為願景訂下的目標化為行動，而

且甘冒風險、不惜自我犧牲來展開行動,從而獲得追隨者的信任。這類領導人能夠一心一意、專注目標,能夠冒著奇險、展開行動,以證明他們對運動、對事業的無私奉獻。他們冒的險愈大,獲得的信任與魅力領導效益也愈大。

魅力型領導人還需要做幾件事:必須指出現狀的技術性等缺失,提出達成目標的策略與非傳統手段,來向追隨者顯示他們曾經或現在,擁有目標領域的相關知識與專業能力。

▶ 救世主與執行長

哈佛商學院教授庫拉納的研究發現,執行長對公司表現造成的衝擊,實際上沒有一般認定的那麼大。根據庫拉納的評估,公司的表現約有 30% 到 40% 取決於產業效應,約有 10% 到 20% 取決於週期性經濟變化,受執行長影響的部分其實只有 10%。庫拉納在所著《尋找企業救世主:對魅力執行長的非理性追求》(*Searching for a Corporate Savior: The Irrational Quest for Charismatic CEOs*)一書中,對商界過於仰賴明星執行長的迷思提出挑戰。

我們在與庫拉納訪談時，首先問他根據他的研究，領導人的重要性在哪裡。

根據你的研究，領導人並不重要？

我並沒有全盤否定領導人的重要性，我的意思是就經濟表現而論，領導人的影響並沒有我們想像中那麼大。

在過去，有關領導學的傳統研究，一般很少涉及對公司表現的分析。對於組織與績效表現的關係這個議題，我們的前輩有比較精密的了解。在過去 20 年來，我們在個別領導與績效成果之間，做了一些很簡化的連結。

對於領導人在組織裡做些什麼這個議題，傳統領導理論的探討也比較精密。但這些理論不能說明 20 年來崛起的自大觀，它們只認定領導人在組織裡扮演的，事實上是一種為參與人創造意義的角色，營利性組織也不例外。

Are you suggesting that leaders simply do not matter?

My intention was to say that when it comes to economic performance, leaders don't matter in the way we understand them to matter, not to dismiss leadership completely out of hand.

Unlike that today, little of the classical research on leadership focuses on how to dissect the performance of companies. Our predecessors had a more sophisticated understanding of the relationship between organizations and performance. Over the last 20 years, we have made simplistic connections between individual leadership and performance outcomes.

Classical leadership theory is also more sophisticated about what leaders do in organizations. It doesn't support the kind of egotistical notion that has emerged over the last 20 years, but rather takes the view that leaders in organizations actually play a role in creating meaning for their participants. That includes for-profit organizations.

——

這麼說，你的看法是，領導人扮演的不是一種膚淺、一種與形象有關的角色，它應該是一種更有深度的東西？

除了公關部門、媒體與執行長們，為了創造魅力形象而積極參與的那種假魅力領導以外，領導還有許多型態。

事實上，領導人會為了追隨者創造條件，讓追隨者從他們的機構中取得意義。領導人這些作為所造成的衝擊，未必能夠立竿見影，雖不能從諸如投資報酬報表中看出

來，但可以在其他各種活動與行動中顯現。

舉例來說，我們經常注意的是，領導人能不能透過自己
的行為與行動來維護組織價值。領導人會做一些結構性
決策，讓追隨者從工作中獲得很大的意義，如領導人會
讓工作有多樣性、有自主性，還能提供反饋等。領導人
會聘用價值觀與使命感與組織一樣的人才，並設法留住
這些人才，以創造價值。有些非常重要的行動類型，在
下一季的財報中是一眼看不出來的。

So are you suggesting that the leader's role is not a superficial, image-related one, but something far deeper?

There is more to leadership than the type of pseudo-charismatic leadership where the PR department, the media, and the CEO actively participate in creating an image of charisma.

In fact, leaders create the conditions for people to derive meaning from their institutions. They do it in a way where the impact is not directly seen in ROI, for example, but rather comes from a variety of other activities and actions.

So, for example, we often focus on a leader's ability to uphold the organization's values through his or her own behaviors and actions. The leader makes architectural decisions that allow people to derive a great deal of meaning from their work—by ensuring that the work has variety and autonomy and provides feedback, for example. The leader creates meaning by paying attention to hiring and retaining individuals who behave in a way that is consistent with the organization's explicit values and purpose. Those are really important types of actions that don't manifest themselves in the next quarter's reports.

——

過去 20 年的領導模式，似乎有許多瑕疵。在今天這個新時代，我們是不是需要一種新的領導模式？

我認為，我們確實需要一種新的領導模式，我稱這種新模式為「制度性領導」（institutional leadership）。

在現代社會，「制度性領導」是絕對不可或缺的要件。我們必須了解，社會上有許多人擁有表面上看不見的才能。許多人不知道他們自己真正的潛能，領導人需要懂得運用這些潛能。

領導人需要了解，他們領導的組織制度，有許多是為了解決社會上某些問題、某些議題而創辦的。領導人或許很有才幹，能夠親自了解並解決這些問題。但他們還必須建立一套制式化程序，將領導效應擴散到組織的每

個角落，讓組織成員在他們不在的時候，也能朝目標邁進。我們要找的，是這一類型的領導力。

也因此，我所謂的「制度性領導」，指的是領導一個組織改變、轉型，賦予它一種與社會價值吻合的使命感與價值觀，讓它超越自身的功利主義功能。

The leadership model of the last 20 years seems to have a lot of flaws. Is a new kind of leadership needed for the modern age?

I think we do need a new model of leadership. I would call it institutional leadership.

Institutional leadership is absolutely critical in contemporary society. We have to understand that many individuals in society have latent talents. People are unaware of their true potential. Leaders need to be able to tap into that potential.

Leaders should understand that many of the institutions they lead were created to solve a certain set of problems and issues in society. Leaders may be quite gifted when it comes to understanding and solving those problems personally. However, they must also create and institutionalize a process by which leadership is distributed throughout the organization, so that it can be carried on in their absence. It is that kind of leadership that we have to find again.

And so what I mean by institutional leadership is changing
and transforming an organization beyond its utilitarian
function—infusing it with a sense of purpose and with values
that are consistent with the larger values of the society.

———

這類未來領導人，必須扮演什麼角色？

未來領導人的任務涉及各種因素，但制度性領導人必須
特別強調的角色，包括尋找人才、訂定目標、強調神聖
價值、鼓勵他人朝這些崇高的目標邁進等。

相較之下，在許多商學院學生眼中，股價才是領導人好
或不好的指標。這樣的眼光不僅非常狹隘，用這種方式
了解領導在商業的重要性，幾近於反諷。我認為，當從
商的人願意質疑最為人信以為真的假設時，商業是一種
高尚的生活方式。

與個人式領導不同的是，商界與商界領導人最後必須面
對的裁判，不僅是股價漲了多少而已，誠實、自制，以
及其他一些經常與制度性領導相提並論的價值，也會成
為他們成敗定位的標準。

What role must these future leaders play?

The task of the leader in the future involves a variety of factors. But institutional leaders, in particular, must find abilities, set goals, reaffirm sacred values, and motivate individuals toward those high values.

In contrast, many business students look at the scorecard of stock price as an indication of whether someone is a good leader or not. That's a very narrow, almost cynical way of understanding the importance of leadership in business. I believe that business is an honorable way of life when people are willing to question the most cherished assumptions.

Ultimately, businesses and business leaders have to be judged not simply by the gain in the stock price, but by qualities such as honor, self-restraint, and the kind of values that are often associated with institutional leadership, as opposed to individualistic leadership.

――――

那麼，我們應該怎麼評估制度性領導人的績效表現？

我們可以問一些這樣的問題：在領導期間，這位領導人增加還是減少了組成分子對組織的信任？在那段期間，就我們期待組織制度能宣揚、代表的那些價值而言，這位領導人能不能身為表率？能不能為系統注入活力，讓

組織制度更能因應挑戰？

So how should we assess the performance of institutional leaders?

By asking questions like: During a person's leadership, did he or she increase or decrease constituents' trust in that institution? During that person's reign, did he or she serve as a symbol for others, in terms of representing the kinds of values that we want our institutions to articulate and represent? Did he or she renew the system in a way that allowed that institution to be better aligned to its challenges?

————

有效領導人究竟做些什麼？

他們倡導權力共享，賦予部屬主動權與責任。他們解決那些使組織癱瘓、令組織無法實現更遠大目標的緊張與衝突。他們開創並找出資源，以利組織運作。他們究竟做什麼很難敘述，領導是一種非常複雜的社會程序。任何類似這樣的討論，都會使領導看起來比實際上有秩序得多。

領導人必須做決定，他們只能憑自己的判斷行事。他們必須知道自己做出的決定，是對的還是錯的。有時他們會修正決定，有時他們會扭轉先前的決定。誤解，是經

常發生的事。

What do effective leaders actually do?

They lead by sharing power, by spreading initiative and responsibility. They resolve tensions and conflicts that paralyze organizations and prevent them from realizing larger objectives. They create and identify resources that allow the group effort to be carried out. It is very difficult to describe. Leadership is a very complex social process. Inevitably, any kind of discussion like this makes it seem more orderly than it really is.

Leaders make decisions; they act on them; they realize that those decisions are right or wrong. In some cases, they revise them; in other cases, they reverse them. Misunderstandings are frequent.

——

追隨者呢？他們在這個領導方程式的位置在哪裡？

領導人與被領導人之間的互動很複雜，了解這個問題很重要。我一直相信好的追隨者能造就好的領導人。而且就若干方式而言，從我們選出的是什麼樣的領導人，就能看出我們的社會像什麼樣。

我們需要討論追隨這方面的失敗。過去幾年來，我們的

討論總不外乎領導方面的失敗，但追隨這方面的失敗仍然需要討論。

19 世紀德國唯心論哲學大師黑格爾（Georg Hegel）曾說，我在這邊稍微改述一下，悲劇不是「惡」戰勝「善」造成的，而是兩個局部的「善」相互迫使對方臣服的後果。我同意他這句話。

What about the followers? Where do they fit into the leadership equation?

It is important to understand the complex interplay between the leaders and those who are led. I always believe that good followers produce good leaders. And in some ways, whom we choose as our leaders tells us a lot about our society.

We should talk about the failures of followership. While we have focused on the failures of leadership in the last couple of years, there still needs to be a discussion about the failures of followership.

I agree with Hegel, who said, and I'm paraphrasing here, tragedy is not the result of evil triumphing over good, but the result of two partial goods trying to impose their will on each other.

——

著眼於近利與具有長期策略的領導之間，似乎仍存在一種緊張。

之所以讓人有情勢緊張的感覺，是因為領導人沒有說清楚，大家都不知道領導人想把組織帶到哪裡去。如果有很明確的宗旨，有一套很清楚的目標，就能讓每個單位、所有相關人等都了解。之後，領導人就可以說明自己為了達成這項宗旨與目標，而在組織裡做成的決定。在這時候，大家都會耐心地聽領導人解釋。

但如果你對某些企業領導人窮追猛打，硬要他們說明組織策略或宗旨，他們很可能只會說一些股東價值、顧客至上之類的陳腔濫調。這類陳腔濫調讓人看不出明確的方向，很難用它們推論出背後的邏輯。好的領導人需要非常明確的敘事能力，想具備這種能力，首先要有很好的環境意識，必須要知道你想達到的是什麼。你要把你的宗旨和目標，清楚、明白地告知股東和資源提供者等人。把事情說清楚，對工作的推動會很有幫助。

There still appears to be a tension between leadership for short-term gain and leadership with a long-term strategy.

The reason it feels like a tension is that often there is no clear articulation of where leaders are trying to take their organizations. If you have a clear purpose and set of goals, everyone can understand it, and it can be articulated to its various constituents. Then a leader can explain the kinds of decisions he or she is making inside the organizations to achieve those goals and that purpose. And people tend to be very patient.

But if you really pushed and pressed certain corporate leaders on what the strategy of their organization is, what its purpose is, and what they are trying to get done, they would have a hard time getting much further than platitudes about shareholder value or pleasing the customer. Those kinds of platitudes don't offer a clear direction, so it is hard to articulate the logic behind them. You need to have very clear articulation, and that requires having a good sense of the environment and knowing what it is you are trying to get done. You need to be able to articulate that to your constituencies, like shareholders or resource providers, and it is that clarity that helps.

———

領導這項議題，今後會面對哪些重大挑戰？

過去幾年來，我們發現敏捷與個人領導，確實擁有一些優勢。今後的挑戰是如何保有其中一些長處，同時想辦法讓我們的領導、我們的組織轉型，進一步響應社會整

體的需求。

領導人必須以促進社會健康繁榮為服務宗旨；要創建一個環境，發揮人們的個別主動精神；要領導一個愈來愈多樣化的世界，抱持多元化的觀點，邁向一個共享的宗旨。

這些都是非常複雜的挑戰，但我們只能接受，不能迴避。我們正在設法保留長處，同時捨棄容易遭來危險與不必要之處。

What are the big challenges ahead concerning leadership?

Trying to figure out how to preserve some of the benefits of nimbleness and individuality that we have seen over the last few years, and at the same time transform our leadership and our institutions to become more responsible toward society at large.

To recall that the ends that leaders serve have to do with things like ensuring a healthy, vibrant society; creating context that takes advantage of people's individual initiatives; and leading an increasingly divergent world, with divergent points of view, toward a common shared purpose.

Those are very complicated challenges. But that is the way things need to go. We are trying to figure out how to preserve the good and at the same time get rid of the dangers and the excesses.

———

▶ 黑暗面走一回

魅力型領導人有說服群眾跟隨他們邁向一個共同目標的能力，他們憑藉這股能力在組織裡成就豐功偉業。然而，他們有時也會濫用憑藉魅力而來的權勢。魅力型領導人一旦步入歧途，後果可能非常嚴重。

1990 年代末期至 2000 年代初期發生的一連串企業犯行，讓人對領導與魅力產生新的疑問。世界通信公司（WorldCom）的伯尼·艾伯斯（Bernie Ebbers）與泰科（Tyco）的丹尼斯·柯茲洛斯基（Dennis Kozlowski）都是魅力型領導人，都曾率領他們的組織走向失敗。促成安隆瓦解的主因，也是魅力加貪婪調成的那杯醉人雞尾酒。直到事發前最後一刻，前安隆執行長傑佛瑞·史基林（Jeffery Skilling）與財務長安德魯·法斯陶（Andrew Fastow），仍在公司集會中魅力四射，讓在場投資人與分析師如醉如痴。一位圈內人甚至說，這些集會不像公司法說會，更像教會的復興布道會。

早在這類事件爆發以前，康格已經在與拉賓德拉·卡農高（Rabindra N. Kanungo）合著的《組織的魅力型領導》（*Charismatic Leadership in Organizations*）一書中，注意

到「領導的陰暗面」。康格與卡農高認為，魅力型領導有正、負兩種形式之分，負面領導人的目標與活動為的是私利，正面形式的魅力型領導人則大公無私。

　　走入黑暗面的魅力型領導人妄自尊大、性喜控制，渴望個人權力與成就，行事詭祕，為讓人就範而不擇手段，而且不講道德倫理。負面魅力型領導人往往選擇能滿足自己私欲的願景，為了達到這些願景，不惜將追隨者的心血、活力與資源摧殘淨盡。凱勒曼在 2004 年出版的《壞領導：它是什麼、怎麼發生，為什麼事關重要》（*Bad Leadership: What It Is, How It Happens, Why It Matters*）一書中指出，有效領導與壞領導兩者未必互不相容，非常有效的領導人仍然可能是壞領導人。

　　壞領導人之所以能夠吸引到追隨者，與好領導人能吸引到追隨者的原因幾乎一樣。壞領導人至少能在一段時間內，提供秩序與結構、安全、簡單與安定。他們能夠栩栩如生地描繪出願景，也能夠組織完成集體工作。大體上，壞領導人可以分為兩類：無效領導人與無德領導人。無效領導人什麼事也幹不成，他們的特徵是「缺乏領導資質、技巧不足，策略構想很差，戰術運用也很糟糕」。無效領導人或許也會訂定一些讓人豔羨的目標，問題是他們缺乏

達成目標的手段。

如果跟在無效領導人背後的，是一群無效的追隨者，那便是一組強大的無效軍力。最有效的追隨者能夠「自行思考、自己安排工作，做好份內的事。他們不斷努力，使自己與企業融為一體，更精益求精、盡心盡力為組織貢獻，與同事合作」。無效的追隨者不具備這一切；他們軟弱，不能獨立行事，也不能或無意為組織盡力、對組織奉獻。

無德領導人雖然知道什麼是對的，但選擇了另一個方向。好的領導一般都是有德的領導。政治學家伯恩斯認為，有德的領導人有 3 種特質：將追隨者的需求置於自身需求之前；體現勇敢與誠實等美德；為謀求公益而行領導。壞領導人可能無效、可能無德，也可能既無效又無德。

▶7 種壞領導人

凱勒曼說，壞領導人有 7 類。「無能領導人」（incompetent leaders）欠缺「維持有效行動的意志或技巧（或者兩者都缺）」。此外，他們可能在許多方面都無

能，例如他們可能很笨、沒有 EQ，行事雜亂無章，或者很懶等。

一般都將彈性視為成功組織不可或缺的要件，懂得調適、能夠因應新情勢，也是成功領導人的關鍵特質。也因此，「僵化領導人」（rigid leaders）一般都是壞領導人。僵化領導之所以出現，是因為領導人與部分（如果不是全部）追隨者缺乏彈性，而且不肯讓步所致。他們或許有能力，但他們欠缺調適力。有些領導人只因過去採取的一些行為途徑曾經帶來成功，就堅持採取同樣的途徑。儘管時過境遷、過去那套成功手段已經不再有效，也不肯調適、通融。這類領導人的例子很多，我們都不陌生。

好的領導人懂得節制；只憑一己情緒行事的領導人是壞領導人。「不知節制的領導人」（intemperate leaders）欠缺自制，在這種情況下，如果追隨者不願或不能進行干預，事情會變得更糟。有些領導人喜歡領導，卻不關心被他們領導的人。這類領導人不在乎追隨者的需求、願望與希望，對於與他們接觸的其他人也漫不經心。

這類自私的領導人真正關心的，只是遂行一己之私。他們一心一意，為的只是達成自己的目標，而且樂意為達

成這些目標而踐踏他人。自私的領導人凡事以自身需求為第一優先，其他一切都是其次。凱勒曼稱這類領導人為「硬心腸領導人」（callous leaders），他們冷血無情，不在乎、不理會其他團體或組織成員的需求與願望。

有些領導人根本就是「惡棍領導人」（crooked leaders），近年來爆發的各式各樣企業醜聞，揭露了一大堆爛蘋果。為什麼有些領導人貪污？千百年來，壞領導人之所以壞，原因幾乎都一樣。凱勒曼說，領導人之所以壞，主要原因就在一個「貪」字：貪財、貪勢，或貪其他一些稀有資源。在愈來愈講求結果的今天，世人用一套標準來評估表現，而表現的好壞決定了報酬。在這種情況下，舞弊的誘惑比過去更大了。行賄與貪污看來不是一種新現象，據說古羅馬的一些皇帝，就是這類詐術的高手。貪腐的領導人如果說謊、欺騙、偷竊、把私利置於公益之先，當然不具備好領導所需的道德要件。

把頭埋在沙裡，或故意視而不見，也不是偉大的領導。碰到發生在團體、組織或他們直接責任區之外的事情，「冷漠型領導人」（insular leaders）便沒有興趣。他們不理會追隨者的健康與福祉，遇事聳聳肩，說一聲「這不關我的事」，就是這類領導人的寫照。此外，還有一

種「邪惡型領導」（evil leadership）。什麼是「邪惡型領導」？用傷害他人的方式取得滿足，是邪惡型領導的一項要件。邪惡型領導人不僅讓他人恐懼，還會想辦法延續他人的苦難。在邪惡型領導個案中，領導人與至少部分追隨者犯下暴行，他們用痛苦做為伸張權力的工具，這種領導一般造成嚴重的心理與肉體創傷。

雖說許多領導人是徹頭徹尾的惡棍，還有許多領導人屬於灰色地帶。有時想判定領導人是好是壞，或者是否漫不經心很難。事實上，同一位領導人可能在不同時間，展現出不同的領導方式。壞領導人是否也值得我們研究？凱勒曼認為，知道壞領導人與壞追隨者的組成要件以後，我們才能設法避開。

▶ 聰明反被聰明誤

當然，就統計數字而言，壞領導在整個領導版圖上只占很小一部分，甚至就政界而言也不例外。但撇開壞領導人不計，好的用心照樣可能為禍。達特茅斯塔克商學院（Tuck School of Business）教授席尼・芬克斯坦（Sydney Finkelstein），是《從輝煌到湮滅》（*Why Smart Executives*

Fail）等多本暢銷書的作者。被芬克斯坦列為研究對象的這些領導人，無不朝氣蓬勃、氣宇不凡，失敗似乎是一件與他們扯不上關係的事。我們在與芬克斯坦訪談時，首先提出的，就是這個價值千萬的問題。

精明的主管，為什麼失敗？

如果深入發掘，你會發現一切有關領導的議題，事實上都是人的問題，都是人怎麼做決定或不做決定的問題。就很大的程度而言，它談的是人怎麼行為。當你經營一家公司的時候，犯錯的許可度通常比在日常生活中的要小得多。

許多領導人不肯承認周圍世界正在迅速改變；他們拖延，把頭埋進沙裡，躲避反饋、特別是負面反饋，安插一堆崇拜他們的人在身邊。這些行為終有一天會反撲，對他們造成重大傷害。

So, why do smart executives fail?

When you dig down into what's behind it all, it's really all about people and how people make decisions or nondecisions. To a large extent, it's about how people behave. When you're running a company, the margin for error is usually a lot narrower than it is in everyday life.

Behaviors like not wanting to acknowledge that the world is changing around you; procrastinating; sticking your head in the sand; avoiding feedback, especially negative feedback; and surrounding yourself with people who think you are great can really come back and bite you.

———

雜誌、書刊中，介紹了那麼多偉大領導人的例子，你對這些例子有什麼看法？

執行長們出了一大堆書，吹噓他們多偉大，對那些沒有做好的事卻隻字不提，讀者對這種現象已經有點厭倦了。

除了這些書以外，還有顧問出的書。這類書籍可能蒐集 10 個不同客戶經驗，而且每個經驗都告訴你，要你採取某一套行動。但他們沒有告訴你，數以百計的其他公司也曾採取同一套行動，最後卻以失敗收場。

就某種程度來看，史蒂夫·賈伯斯就是一個好例子。每當我談到有關謙卑、開放胸襟與調適的議題時，聽我說這些話的人每兩、三個人，總會有一人對我說：「你說的這些話都有道理，但史蒂夫·賈伯斯與你說的這些，

一點也扯不上關係。我看過他的書。」史蒂夫還在世時，情況尤其如此。

這話說的不錯。賈伯斯根本不知謙卑為何物，也從來不肯自我調適。但如果你想根據史蒂夫・賈伯斯建立一套領導理論，你能夠取用的樣本只有一個。而且我敢說，採取賈伯斯式管理與領導風格的人，1 百人裡面有 99 人會以失敗收場。我所以敢這麼鐵口，是因為我觀察到的那些執行長，都是敗在這類管理與領導風格上。

我們很容易陷入只觀察一個例外，或只觀察兩、三個偉大成功故事的陷阱。因為這類例外與成功故事家喻戶曉，大家都知道。但這麼做的後果是，你忽略了真正行動的所在。

如果真正關心組織如何運作、領導人如何思考、如何做決定等問題，你需要蒐集變化多得多的各式樣本。觀察問題不僅要注意對的事，也要注意錯的事，這點很重要。

What about all the examples of great leaders we get to read about in articles and books?

People get a little get tired of seeing all these books from CEOs about how great they were, ignoring all the things that didn't go so well.

And then you have the consultants' books that look at maybe 10 different client experiences, all of which tell you to take a certain course of action. But what they miss is the hundreds of other companies that may have failed doing exactly the same thing.

In a way, a modern example of that is Steve Jobs. When people hear me and some of the things I talk about related to humility, open-mindedness, and adaptability, one out of every two or three times somebody will say, "I hear what you're saying, but Steve Jobs didn't do any of that; I read the book." This was especially true when Steve was still alive.

And that's right. He didn't do any of that. But if you want to create a theory of leadership based on Steve Jobs, you're going to have a sample size of one. And I'm quite convinced that 99 out of 100 people that adopt that management and leadership style are going to fail. The reason I'm quite convinced of this is that it's similar to the failing CEOs I looked at.

So it's easy to fall into the trap of looking at an exception, or a couple of great success stories, because they get a lot of publicity and people read about it. But you miss out on where the real action is.

If you really care about how organizations work, how people make decisions, and how leaders think, you need to have a much more diverse sample. That's the importance of looking at not just what's right, but what's wrong.

———

領導人除了應該從好領導人身上學習以外，還應該從壞領導人身上學習嗎？

我的書剛出版的時候，美國《商業周刊》（*Businessweek*）寫了一篇討論我的專文，不是書評，是專文。談到我的理論，也談到這個主題。

令這篇專文作者感到匪夷所思的是，在商學院這種地方，大家全心注意的是怎麼做最好，不是怎麼做不好，而我卻在這裡大談領導出的亂子。

事實上，想要了解問題，好、壞兩面都要注意。許多研究成果已經證實，在許多領域，多樣化或多元性的觀點或經驗，能把人訓練得更好。

Leaders should learn from the bad as well as the good?

When my book first came out, *Businessweek* wrote an article—not a review of the book, but an article—about me, how I teach, and the subject.

What the author thought was so intriguing was that I was teaching about what goes wrong in places like business schools, where people spend all their time looking at the best practices rather than the worst practices.

In fact, you need a combination of both. There is a fair amount of research that shows that diversity or diverse perspectives or variance in experience leads to the better training of individuals in a lot of walks of life.

———

依你看，領導人願意虛心學習嗎？你會不會覺得他們比較重視現實？

這個問題，隨產業不同而不同。一般的情況是，許多人雖然知道，領導人如果採取那種唯我獨尊的執行長式世界觀，可能會出亂子，但想因此而有所行動很難。如果容易，每個人都可以做到了。

想在組織裡晉升頂尖，你必須擁有巨大的自我意識，而且不斷做出必要犧牲，例如犧牲個人生活、犧牲家庭生活等，來力爭上游。

某些人格類型就擁有這種巨型自我，能讓人這麼犧牲。

但同時，這世上也有像我這樣的人，會不斷地提出告誡：「要讓其他人參與，要聽取其他人的觀點，不要驟下結論，胸襟要開闊，要考慮世事多變，要做好改變過去行事慣例的準備。」不過，想讓他們接受這個訊息，很難。真的很難。

Are leaders learning, do you think? Is there a sense that they are more grounded in reality?

It varies across industries. In general, while there is much more awareness of what can go wrong when a leader adopts the imperial CEO view of the world, it's hard to act upon that awareness. If it were easy, everybody would do it.

To get to the top of an organization, you've got to have this gigantic ego that keeps you going and allows you to make the sacrifices required, in terms of your personal life and your family, for example.

A certain personality type can make that kind of sacrifice and has a big ego. At the same time, there are people like me and others who are saying, "Bring in other people, listen to other people's points of view, don't jump to conclusions, be open-minded, think about how the world's changing, and be prepared to change what you've done in the past." Boy, that's a tough message. It's difficult.

──

某些類型公司的領導人，在這方面是不是做得比較好？

我認為，新創企業幾乎像是天性使然一樣，比較容易接受這個訊息。近年來，「轉軸」（pivot）這個字，已經成為有關策略討論的一個熱門字眼：新創企業能夠從一個軸點旋轉到另一軸點。

在新創企業中，你會放棄行不通的商業模式，採取你希望能行得通的新商業模式。事情就這麼簡單，公司還在草創階段，就算失敗也沒多大損失，所以你進行改革，採取新做法。

但是，要一家大公司改變過去的做法，改變商業模式採取新模式……這種例子並不多。手機產業就是一個有趣的例子，最先是摩托羅拉（Motorola）獨領風騷，之後諾基亞（Nokia）擊敗摩托羅拉，然後你看到 RIM 公司（Research in Motion）與黑莓機（BlackBerry）超越諾基亞。接著，蘋果 iPhone 崛起，現在又有三星（Samsung）大放光明。

一波波後浪不斷吞噬前浪，巨型市場占有率與品牌霸權

不斷易手：手機產業的故事，令人看得瞠目結舌。

在這類產業，想生存，就必須懂得調適、必須保持開放的胸襟才行。但要一個人願意拋棄過去慣用的做法，並不容易。

Do leaders in certain types of companies do this stuff better?

I think entrepreneurial companies tend to be more embracing of the message, almost by nature. One of the big words in strategy lately has been pivot: they pivot from one spot to another.

In an entrepreneurial company, that just means that you're changing your business model from something that didn't work to something that you hope is going to work. There's not that much to lose at that stage, so you do it.

For a big company to change what it's done and its portfolio of businesses into something else . . . well, there are not a lot of examples of that. The cell phone industry is an interesting example. You go from Motorola to where Nokia beat Motorola. Then you go past Nokia to Research in Motion and the BlackBerry. Then you go to the Apple iPhone, and now to Samsung.

It's an incredible story in that industry, where huge market share and huge brand name power are eclipsed by a subsequent player.

So there's a requirement that you be adaptable, open, and adjusting. Being willing to throw out what you've done in the past is not easy.

——

當你環遊世界、討論《從輝煌到湮滅》這本書的時候，有沒有碰上什麼不同的文化反應？

基本上，我在美國與英國碰到的反應是：「你說的有道理，既然你這麼說，我也願意買帳，但或許你應該找我的老闆、執行長、董事長與執行主席談一談。」

換句話說，我得到的是一種推諉責任的答覆，而我不接受這樣的答覆。當然，他們說這話，也並非全無道理。不過，如果你身為中高階經理人，難道就不能採取行動，改變你的世界、你的團隊、你的部門、你的辦公室嗎？

你當然有這份能力。你只要真的認為，我建議的這許多原則與構想有道理，值得你採納，就可以在你本身的領域進行改革。

如果改革之後，你取得較好的成果，這本來就是改革的前提，你就可以向他人訴說一個非常有趣的故事——而每

個人都喜歡聽好故事，不喜歡聽壞故事。

就這樣，有關這個構想的談論，開始緩慢但穩定地傳開了。

Are there different cultural responses when you travel around the world talking about *Why Smart Executives Fail*?

The primary tendency I see in the United States and the United Kingdom is, "Well, it's fine, you're talking to me and I buy it, but maybe you need to talk to my CEO, chairman, and executive chairman."

In other words, I get a pass-the-buck type of answer, which I never accept. Of course, there's some element of truth to that, but if you're a middle or upper-level manager, don't you have the ability to change your world, your team, your department, your office?

The answer is yes, of course you do. If you really think that it makes sense to adopt the various principles and ideas that I recommend, then you can change what you're doing in your own area.

If it's going to help you get better results, which is the whole premise, then you have a really interesting story to tell—and everyone prefers to hear good stories rather than bad stories.

And so slowly but surely the word starts to spread about this idea.

———

這麼說，確實有文化差距囉？

當然有。如果你到亞洲，除了中國之外，一般對資深領導人都非常尊重。亞洲有比我們西方社會更悠久的父權文化史。此外，在香港與新加坡，也有一些跨足許多產業而且非常成功的億萬富豪。

所以說，就某個層面而言，想在這些環境中傳播這項訊息，會比較困難。但就另一個層面而言，當我觀察這本書各種語言的譯本時，我發現亞洲地區出版的翻譯版本比其他地區都多，而且原因不只是因為亞洲的國家比較多而已。亞洲地區的譯本，不僅出現得比較快、比較早，發行量也比較大。

So there are cultural differences?

To be sure. When you get into Asia, with perhaps the only exception being China, there's this deference to senior leaders. There's a history of paternalistic culture, a little bit more so than we have in Western society. Also, in Hong Kong and Singapore, there are these billionaires who are in so many different businesses and have been extremely successful.

So at one level, in those environments, it tends to be tougher to get the message across; on another level, when I look at all the international translations of the book, there are more in Asia than anywhere else, and it's not just because there are more countries. They were faster, sooner, and greater in number.

———

在你所著的《Think Again：避開錯誤決策的 4 個陷阱》（*Think Again*）這本書裡——與安德魯·坎貝爾（Andrew Campbell）和喬·懷海德（Jo Whitehead）合著——自我認知似乎是一個重要主題。

《*Think Again*》主要談的是決策，書中探討了一些人腦如何運作、我們怎麼處理資訊，特別是我們怎麼決策等大層面的問題。

我談到自我認知。每當我以顧問身分與一位執行長或高階主管共事時，自我認知的議題經常出現在我們的談話中，而對方的自我認知程度之高，也往往令我稱奇。對我而言，自我認知絕對是一種最有力的領導能力。如果詳細探討，自我認知其實是一種非常情緒、非常本能的東西。我稱它為一種領導能力，為的是讓它看起來比較實際。

人腦有一部分，能讓我們憑直覺與本能行事。但我們如果凡事只憑本能衝動，一定會惹來許多麻煩，而自我認知可以大幅替我們解套。愈是了解自己如何思考、有什麼行為舉止及偏見的人，就愈能掙脫這種直覺、本能的束縛。那有可能會讓你惹出很多麻煩，所以自我認知是重要差別，決策前再想一下。

Self-awareness seems to be an important theme in your book *Think Again* (coauthored with Andrew Campbell and Jo Whitehead).

Think Again is very much a book about decision making, and it gets into the micro dimensions of how people's brains work, how we process information, and, specifically, how we make decisions.

I talk about self-awareness. Working in a consulting capacity with a CEO or senior executive, the extent to which that person is self-aware is really remarkable; it comes out in a conversation so often. To me, it's really one of the most powerful leadership capabilities. That's how I label it to make it seem more practical to people, because self-awareness is a very touchy-feely type of idea once you get right down to it. But I call it a leadership capability.

The more anyone knows about how he or she thinks and behaves, his or her own biases, the less likely that person is to become a slave to that part of the brain where we just do what our gut instinct tells us to do. That can get you into a lot of trouble, so self-awareness is a big differentiator. Think again.

———

這麼說，自我認知與理智性的誠實，應該都與真誠有關係了。

這些自然都是相互重疊的概念。對我來說，理智性的誠實稍微多了一些外向的專注性，因為你觀察的是外在世界，想的是外在世界，設法面對的也是外在世界。而自我認知卻內向專注得多。不過，它們都與真誠有關。

We suppose that self-awareness and intellectual honesty are connected with authenticity.

These are certainly ideas that overlap. For me, intellectual honesty is a little bit more outward-focused, because you're looking at the world out there, thinking about it, and trying to face up to it, and self-awareness is much more internally focused. But they are related to authenticity.

———

第 6 章

追隨者

哈佛大學公共領導講座教授
芭芭拉‧凱勒曼
策略大師
與談人 │ **蓋瑞 ‧ 哈默爾**

早期管理思想家瑪麗・派克・芙麗特（Mary Parker Follett, 1868-1933年）曾說：「領導力的界定，不在於權力的運用，而在於能讓那些接受領導的人增加多少權力感。」

被譽為「管理理論之母」的芙麗特思考與眾不同，早期大多數研究領導的學者，關心的只是領導人──領導人有什麼特質、行為或風格等。雖然也有幾位學者探討了領導的另一個層面：領導人與被領導者之間的關係；畢竟，若是沒有人追隨，領導也不可能存在。學者們在探討追隨者扮演什麼角色的問題之後，發現在領導的塑造過程中，領導人與追隨者之間的互動，無論結果是好或壞，對領導力的影響很大。

▶ 追隨者也可分為 4 種

早期深入追隨者領域進行探討的學者中，有一位哈佛商學院教授亞伯拉罕・薩萊斯尼克（Abraham Zaleznik, 1924-2011 年），他也是一位著名的社會心理學家與領導學專家，曾於 1965 年在《哈佛商業評論》發表一篇〈部屬動能〉（The Dynamics of Subordinacy）。薩萊斯尼克從佛洛伊

德派心理分析的角度觀察這個議題，認為追隨者可以分為積極或消極，服從型或性好控制型。

　　積極的追隨者要的是投入、主動和參與，會在領導人與追隨者的互動過程中扮演積極的角色。消極的追隨者則樂於隱身在後，讓領導人在前方衝鋒陷陣，他們則坐待事情發生。在其他層面上，性好控制型的追隨者會毫不猶豫地投入一場意志之戰，意圖控制他們的老闆。反之，服從型的追隨者則樂於接受領導人的意志，願意聽命於人。

　　薩萊斯尼克以這許多層面為根據，將追隨者分為 4 大類型。既積極、又好控制的追隨者，列為「衝動型」（impulsive）。衝動型追隨者凡事自主、自動自發，很有叛逆性，就算在受人領導的時候，也會想辦法為自己訂定方向，設法帶頭領導。同時，他們也很有勇氣，很敢冒險。至於性好控制但消極的追隨者，雖然也喜歡自主，喜歡支配他們的領導人，但會因此產生一種罪惡感而有所收斂；薩萊斯尼克將這類追隨者稱為「強制型」（compulsive）。

　　積極但服從型的追隨者，樂於聽命於領導人，即使這些命令難以辦到也在所不計；薩萊斯尼克將這類追隨者稱為「受虐型」（masochistic）。最後，還有一類稱為「退避

型」（withdrawn）的追隨者，他們對身邊發生的職場大小事務不聞不問、毫不在乎，只求保住工作就好，其他事情能躲就躲——大多數在組織工作的人，對這類追隨者想必不會陌生。

▶ 向上管理，搞好你與老闆的關係

有領導資質當然是一件好事，但管理好與上司的關係，也能讓你獲益匪淺。哈佛商學院教授約翰‧科特（John Kotter）與約翰‧賈巴洛（John Gabarro）研究的，就是這種領導人與追隨者的關係。兩人認為，所謂「管理你的老闆」，指的就是部屬應該注意他們與上司關係的品質。雖然兩人承認「管理老闆」這話，聽起來似乎有些古怪，但無論對組織或對當事人而言，這麼做都合情合理。畢竟，並非每個老闆都是好領導人，追隨者有時需要在領導人與追隨者關係上下一番工夫，使這個組合能夠發揮出最大的效益。此外，一個善於和部屬互動的卓越領導人，在面對自己的老闆時，未必能做得一樣好。

搞不好與上司的關係，可能會讓你損失慘重。科特與賈巴洛指出，在最糟的狀況發生時，搞不好與上司的關係

可能讓你丟掉工作，甚至讓你身敗名裂。我們時常犯的一項錯誤，便是誤解老闆與部屬關係的本質。這是一種相互依存的關係，領導人與被領導者彼此依賴。不過，做部屬的，往往不願承認自己對老闆的依賴有多深。這是一種短視，特別是當做部屬的需要老闆為他引薦，讓他與組織內其他部門取得聯繫或相關資源時，這種看法尤其不智。做部屬的往往認為，身為老闆當然應該知道要完成任務就得提供什麼幫助與支援。

同時，部屬也時常未能察覺領導人對他們的依賴有多深。領導人也是有感情的人（大多數領導人是這樣的），追隨者的行動無論在專業或個人層面上，都會影響到領導人。如果追隨者的表現不佳，影響到領導人的成績，也會對領導人的職涯造成負面衝擊。領導人依賴追隨者，希望追隨者都能夠誠實、可靠、開明、合作。

科特與賈巴洛說，基於這些理由，在領導人與追隨者關係中的雙方，都「需要好好地了解對方與你自己，特別是在有關長處、弱點、工作風格與需求等議題上尤其如此」。有了這些了解以後，他們需要「運用這些資訊，來建立、管理一種健全的工作關係──一種能與兩人的工作風格與資產共容的關係，一種相互期許、能夠滿足對方最關

鍵性需求的關係」。

　　當然，想深入了解老闆的心理，很難。誰沒有這樣的感嘆：「唉，真不知道老闆心裡到底在想什麼？」完全誤解老闆的例子，在職場生活中更是屢見不鮮。但如果你真的想跟老闆建立一種有生產力的關係，就必須徹底下定決心，深入發掘老闆的心理。科特與賈巴洛說，追隨者至少需要了解老闆的目標與壓力、長處與短處，並應該時時刻刻注意領導人的行為，這樣才能預見行為的轉變。

　　就某些方面而言，了解自己是比較簡單的事。進入自己的內心深處，比進入他人的內心深處要容易得多。想判定自己的長處、弱點、目標與工作風格，應該不至於太難；難的是一旦有必要，你得採取行動。就與老闆之間的互動而言，可能包括了各式各樣的反應，其中有兩個極端：一是反依賴行為，另一是過度依賴行為。有反依賴傾向的人，往往對老闆能夠支使他們的權威深惡痛絕，因此常與老闆起爭執、表現得很叛逆。他們可能為自己受到的束縛表示不滿，可能為抗爭而抗爭，還可能將老闆視為一個不得不容忍的敵人，甚或是一個必須擊敗的對手。一旦碰上一位專制、獨裁的老闆，這種關係可能造成非常惡劣、難堪的後果。

　　相較之下，有過度依賴傾向的人，往往極度卑躬屈膝、唯命是從。即使知道老闆做的決定很差，他們也會遵照辦理、毫無怨言；即使事情有辯論與討論的必要，他們也無意表示意見。過度依賴的主管，往往會「將老闆視為無所不知的父母。當父母的，知道得當然最多，當然應該為他們的職涯前途負責，也當然應該培訓他們，讓他們知道一切需要知道的事。此外，也會保護他們，不讓那些野心過大的同事欺負他們」。但無論是反依賴或過度依賴這兩個極端，都不健康、也都沒有生產性。雖然這些行為可能早已在當事者身上根深柢固、難以改變，但對這類行為有所認知，將能有效因應。

　　一旦你了解自己和老闆的內心運作，就可能開始發展、維持一種對雙方皆有利的工作關係，並在自己的工作習慣中加入老闆喜歡的元素。舉例來說，你老闆喜歡的是書面報告，還是口頭報告、順便聊一下天？在需要做決定的時候，你老闆喜歡親自參與，還是喜歡授權給部屬，只在必要時才聽取報告？你老闆是否很喜歡知道部屬都在做些什麼事？

　　聰明的部屬會有效溝通、有效管理期望，寧可傳遞過多資訊，也不要傳遞過少資訊。身為部屬，如果能夠做到

誠實、可靠，再加上懂得「老闆時間寶貴、不容浪費」，
那麼要建立一種互惠的領導人與追隨者關係應該不難。

▶ 追隨者的素質

羅伯‧凱利（Robert Kelley）在 1988 年的一篇論文中指
出：「我們為了尋找更好的領導人而費盡心血，卻也因此
疏忽了領導人要領導的那些人。組織的成敗，部分取決於
領導人領導得好不好，但部分也取決於被領導的人追隨得
好不好。」

大多數人花在追隨上的時間比他們花在領導上的時間
多；有鑒於此，凱利認為追隨者與追隨的行動值得我們注
意。好與壞的追隨者之間，有什麼不同？凱利說：「有效
追隨者與無效追隨者之間的差異，就在於熱情與智慧。有
效追隨者無須表揚，也能夠自動自發、追求組織目標。」

對組織不滿的員工，可能是個壞追隨者，這道理不難
理解。但對組織同樣忠誠的員工，表現的追隨品質為什麼
也有好有壞？為了答覆這個問題，凱利觀察造成有效與無
效追隨的行為。他發現兩個造成有效與無效追隨差距的行

為層面，並且根據這兩個層面，構築了一個追隨行為型態模式。凱利提出的這兩個層面，都以追隨者思考與行動方式為重點——追隨者會自行思考，還是指望領導人替他們思考？他們會積極參與關係，為工作注入正面活力，還是行事消極，只有負面回應？

▶ 你在座標上的哪一區？

凱利建了一個 2×2 的座標，中間有一個跨越四個區塊的圓圈，上層兩個區塊是獨立性思考與批判性思考，下層兩個區塊是依賴性思考與非批判性思考。座標右側的兩個區塊是積極參與，左側是消極參與。根據這個模式，追隨者可分為 5 種類型。左下角是「羊」（sheep）：他們既消極，又欠缺批判性思考，指望老闆為他們設想，等著老闆督促他們行動。把任務交給「羊」來做，他們雖然會完成任務，但之後就停了下來。他們不會採取主動，也不想負什麼責任。儘管他們聽話，不會與領導人爭論，但與「羊」共事仍然很難，因為領導人必須不斷地交代他們工作、不斷照顧他們。

在座標的右下角，是欠缺批判性思考但行事積極的

「遵命先生」（yes-people）。遵命先生會把交代的任務做好，而且在任務完成以後還會回來找老闆，看下一步該做些什麼。他們非常主動，但只會聽命行事，沒有自己的創意與主見。缺乏自信的老闆，一般喜歡這類型的部屬，因為他們不會替老闆惹麻煩，而且很勤勞。

在座標的左上角，是「疏離派」（alienated）。儘管他們很能自行思考，但很多疑，往往為組織行動帶來負面效應與負面景觀。疏離派在行動上很被動，總是能找出不面對新挑戰、不改革的理由。他們不思進取，對此刻正在發生的一切，也都心存懷疑。但凱利認為，疏離派對自己的行為不會這麼想，他們認為自己鶴立雞群，也認為自己是唯一敢挑戰領導人的追隨者。

至於位於座標中心、在每個區塊都占了一角的那個圓圈，代表的則是「務實派」（pragmatists）。他們是天生的生存者，喜歡守在一邊，靜待事情發展。火車如果出發，他們不會是第一個上車的旅客（免得這班車最後哪裡也去不了），但他們也不會是最後上車的。沒有人能指控他們不參與、不關心。凱利說，務實派生存高手的座右銘，就是「安全比後悔好」。

最後，在座標右上角的，是「明星」（star）追隨者。明星追隨者有獨立精神，能夠批判性思考，行事積極，能為自己與老闆的關係與在組織的工作帶來正面活力。這類追隨者自動自發，對老闆說的一切不會照單全收，懂得運用批判性思考來判斷。如果不同意領導人的決定，他們會提出質疑，並與領導人討論，提出替代解決辦法。明星追隨者或許看起來像領導人一樣，但他們是追隨者，而且有許多人非常喜歡扮演這種極有效的追隨者角色。

▶ 跟隨他人的 7 種動機

除了建立追隨者行為模式以外，凱利還認為有效追隨者具有下列資質：善於自我管理；除了個人興趣之外，還會非常熱情地獻身某個目標、組織或理念；有能力，也非常有技巧；勇敢，而且誠實。經過觀察，凱利發現想當追隨者，有 7 種可能的途徑，每一種途徑各有動機。

「學徒」（apprentice）是追隨者，但他們希望有一天能成為領導人。「弟子」（disciple）喜歡自己的領導人，希望有一天能像這位領導人一樣，而且想和這位領導人在一起。「門生」（mentee）關心的，主要是個人發展，未必涉

及領導。「戰友」（comrade）喜歡與一群人，為一項共同目標一起全力投入的經驗。「保皇黨」（loyalist）因為情緒上的承諾，而願意追隨領導人。「夢想家」（dreamer）有一個當追隨者的夢。還有一些人只為了喜歡追隨者的「生活方式」（lifeway），而願意當一個追隨者。

學者對領導人與追隨者之間的心理關係愈來愈注意，於是有人提出一個問題：為什麼追隨者願意追隨某位領導人，不願追隨另一位領導人？美國人類學者與心理分析師麥考比便針對這個問題，提出一套從心理學角度出發的解釋。麥考比認為，領導人與追隨者之間的心理關係，是一種「感情轉移」（transference）。

「感情轉移」是佛洛伊德提出的一項概念，意指過去關係的情感與經驗（往往是父母與子女的關係），會轉移到現在的關係。佛洛伊德用這項概念，來解釋病人對他的愛慕。根據這項解釋，如果員工認定老闆對他們的關愛，就像父母關愛子女一樣，便會更努力來取悅老闆。這樣的情勢會不斷地持續，直到某天員工能夠認清現實，掙脫這種烏托邦式的關係為止。直到那一天，員工會發現自己身在職場，老闆不是父母。

▶ 領導學的新興議題：追隨者

　　凱勒曼很了解追隨的力量，對凱勒曼而言，對領導的興趣是人類生活的一部分。她說：「我們文化的每一部分，從政治到宗教，甚至是個人關係，都離不開領導。」早在領導成為學術界熱門議題很久以前，凱勒曼就已經迷上這個議題。她 1970 年代初期從耶魯大學政治學系畢業時，博士論文討論的，就是西德總理威利‧布朗特（Willy Brandt）。

　　事實證明，凱勒曼對領導議題的熱情歷久不衰。今天，凱勒曼是哈佛大學甘迺迪政府學院詹姆斯‧麥格雷戈‧伯恩斯公共領導講座教授，也是甘迺迪學院公共領導中心創辦人。當其他人對領導英雄吟誦不已之際，她在 2008 年出版《追隨力：追隨者如何創造改變，如何改變領導人》（*Followership: How Followers Are Creating Change and Changing Leaders*），提出權威的反制論點。她與我們談到，為什麼追隨很重要。

　　在領導這項議題上，我們是一直以來就找錯了地方，還是說我們只注意到一半，忽略了另一半？

我不能說我們找錯了地方，但我確實認為，在過去 25 年來，我們有關管理的書，太過偏重領導人。我不能說，觀察領導人是個錯誤。我要說的是，只看領導人卻完全不理會追隨者，不僅對我們有關領導與管理的思考方式有害，對我們的領導與管理實踐也有害。

Have we been looking at the wrong part of the leadership equation, or have we just neglected one-half of it?

I wouldn't say we've been looking at the wrong part, although I do believe that management books have become far too leader-centric in the last 25 years or so. I would not say that it's a mistake to look at leaders. What I am claiming is that by looking only at leaders and ignoring followers entirely, we're doing a disservice not only to the way we think about leadership and management, but also to the way we practice it.

——

「追隨者」（follower）這個字詞，有「像羊一樣馴服」的含意，這會不會也是造成問題的部分原因？

誠如你所說，「追隨者」這個字詞，往往與「羊」畫上等號，我們之所以不願意討論追隨者，這是原因之一。舉例來說，我可以向你保證，哈佛甘迺迪學院的課程只

談如何教育領導人，對追隨者的教育問題隻字不提。誰
願意當一隻羊？每個人都希望自己偉大、重要、成功，
沒有人希望自己低下卑微、不受人重視。但現在，這種
價值判斷已經出現重大修正。

**The word follower has connotations of sheeplike behavior. Is
that part of the problem?**

As you say, the word follower is often equated with being a
sheep, and this is one of the several reasons that we stay away
from it. The Harvard Kennedy School, for example, is all about
educating leaders, and I can assure you that you won't hear a
word about educating followers. Who wants to be a sheep?
Everybody wants to be big and important and successful, and
nobody wants to be lowly and ignored. However, as we speak,
this value judgment is undergoing a profound revision.

——

那麼，為什麼現在會出現這種修正？

首先，就歷史意義而言，追隨者的地位，一直比一般認定
的更加重要。其次，也是因為我們的社會變了，最明顯的
是文化與科技出現的變化。基於這些原因，今天的追隨者
比過去任何時代都重要，領導人的重要性正在減少。

Why is that happening now?

First of all, historically, followers have always been more important than is generally presumed. And second, because of changes in society, two of the most obvious being changes in culture and changes in technology. So, followers are more important now than ever before, and leaders are less important.

———

追隨者的定義，是一個問題。妳怎麼判定一個人是不是追隨者？

沒錯，如何為追隨者下定義，確實很複雜。我來給你舉一個例子，在美國，特別是在小布希（George W. Bush）政府主政初期，許多人認為小布希總統是副總統迪克・錢尼（Dick Cheney）的傀儡，錢尼才是幕後操縱人。根據這種說法，你可以將小布希視為錢尼的追隨者。

為了避免這類複雜，我將追隨者定位為位階較低的人。換句話說，在團體與組織裡，他們是上司的部屬。所謂追隨，指的只是位階較低者對於位階較高者的反應。追隨者一般都會遵照位階較高者的意願行事，但絕非一直這樣。

追隨者有時候會抗拒、不理會他們的領導人。所以，我
們不能認定追隨者就會自動追隨，特別是在今天這個時
代，尤其如此。我們可以假定的是，追隨者位階較低。

One of the problems with followership is its definition. How do you decide when someone is a follower?

Yes, it is complicated. I'll give you an example. In the United States, especially in the early years of the Bush administration, many people felt that the president of the United States, George W. Bush, was the puppet of the vice president, Dick Cheney, who was the puppeteer.

So, in that sense, you can argue that Bush was a follower of Cheney.

To avoid those sorts of complexities, I define followers very clearly as those who are of lower rank. In other words, they are subordinates to their superiors in groups and organizations. And followership is simply the reaction of those of lower rank to those of higher rank. Now, followers generally go along with those of higher rank, but they by no means always do so.

Sometimes followers resist, and sometimes followers ignore their leaders. So we cannot presume, especially in this day and age, that followers will automatically follow. What we can presume is that they are those of lower rank.

——

這麼說，追隨者的定義，以一個人在組織中的階層為根據？

是的。在一家例如醫院的組織中，階層一般很嚴格。在其他的組織裡，例如在一個國家，則比較沒那麼嚴格。

So that definition is based on where someone is in the hierarchy?

It is. In an organization such as a hospital, for example, the hierarchy is generally strict. In other organizations, such as a nation, it is less so.

——

所以說，並非所有追隨者，都生來平等？

沒錯。而且不只是所有追隨者，並非生來平等而已，我們還犯了一個把他們都混為一談的大錯。我們花了許多時間討論我們的領導人，但沒有花任何時間觀察我們的追隨者。看到這世上竟有這麼多各式各樣的追隨者，我認為絕對有必要將他們分門別類，絕對有必要根據他們的活動與承諾程度，將他們分門別類。

於是，我將追隨者分為 5 類。位於最極端的，是我所

謂「隔離型」（isolate），這類追隨者與他們隸屬的
團體或組織完全隔絕或脫節。第二，是「旁觀型」
（bystander），這類追隨者未必脫節，但他們基於一套理
由選擇不參與，於是他們袖手旁觀，什麼也不做。這一
型的極端例子，曾出現在納粹統治下的德國。當時許多
德國人並不同意希特勒的作為，但他們袖手旁觀，什麼
也不做。第三種追隨者是「參與型」（participant），這類
追隨者通常會支持領導人，但並非沒有例外。參與型追
隨者通常在組織中很活躍，但不會對組織過度奉獻。

So not all followers are created equal?

Yes. Not only are not all followers created equal, but we have
made a big mistake by lumping them all together. We spend a lot
of time dissecting our leaders, but no time looking at followers.
When I looked at all the different types of followers, I thought
it was absolutely necessary to break them down, and absolutely
necessary to break them down according to their level of activity
and commitment.

So, there are five types of follower. At one end of the spectrum,
there are what I call isolates, who are completely alienated or
detached from the group or organization to which they belong.
Then there are bystanders, who are not necessarily detached,

but who choose, for whatever set of reasons, not to participate, so they stand by and do nothing. An extreme example is what happened in Nazi Germany, where many people did not agree with Hitler but stood by and did nothing. The third type of follower is the participant, who generally, but not always, supports the leader. Participants are generally active in the organization, but not overcommitted to it.

————

第四類呢？

這 5 類中的第四類，是有時稍嫌莽撞的「積極派」（activist）。積極派型的追隨者非常活躍，對組織或對領導人非常奉獻。他們有時也會全力以赴、擊敗領導人，雖然他們有時並不支持領導人，但在改變領導的事情上卻非常消極。所以，他們未必支持領導人。最後一型的追隨者，是我所謂的「死硬派」（diehard）。這類追隨者，真的會隨時準備為他們篤信的理念或對象效死，為他們誓必推翻的事物犧牲生命。

And the fourth type?

The fourth of the five types is the activist, who is gung ho in some way. These people are very active and very committed to either the organization or the leader, or they might be committed to upending the leader. Sometimes people do not support the leader but are very passive in trying to change the leadership. So, again, this is not necessarily supportive of the leader. And the last kind of follower is a group that I call the diehards. Those in that category, and I do mean this literally, are ready, willing, and able to die for the cause or the person in whom they believe, or whom they wish at all costs to overturn.

———

最後這一型的很極端，在政治之類的情勢中很管用？

沒錯，還有在軍事上，也很能派上用場。

The last group in particular are the extreme—that could have an application in political situations and so on?

Yes, and in the military.

———

回到公共組織的層面上。這幾類追隨者，對領導人有什麼影響？

在今天這個時代，追隨者如果想讓領導人日子不好過，要比過去簡單得太多。所以領導人為了自己的利益，也

應該設法善用追隨者的潛能。

不能認認真真了解部屬的領導人，會錯失明智領導的大好良機。你不能認定所有部屬都一樣。你需要將他們分門別類，才能針對特定目標，進行適當領導。這有點像是推銷一件產品；首先你得選定推銷對象，不能不分青紅皂白、只採用同樣的推銷手段，因為你面對的人事實上非常不同。

所以，對領導人來說這點很重要：要能夠認清在現在這個時代，他們其實很容易受到追隨者的影響。第二，以正面一點的角度來說，他們必須要能夠將自己的產品推銷出去——他們想達成的目標——給心中預定的對象。

對追隨者而言，這點也非常重要：我們大部分的人，就算幾乎已經爬上頂層，由於仍然得面對上司，所以都是追隨者。特別是在像醫院這樣的組織，傳統上非常重視組織階級，在美國境內的醫院尤其這樣。現在就是追隨者與部屬，進一步了解彼此能夠達成什麼成就的時候了。在許多傳統上階級嚴明、讓追隨者膽怯的組織，只要懂得某些戰術與策略，以及被聆聽的方式，想開始進一步了解並不難。

Coming back to a public organization, what are the implications of those different types of followers for leaders?

In this day and age, followers can make things difficult for leaders much more readily than they used to be able to do. So it is in the interest of leaders to harness the potential power of followers.

Leaders who don't pay serious attention to those who are their subordinates miss out on an enormous opportunity to lead wisely and well. You can't presume that all your subordinates are the same. You need to be able to distinguish among and between them, in order to lead appropriately for the particular audience you're trying to target. It's a bit like selling a product; you need to target your audience instead of simply trying to sell it in the same way to people who are in fact very different.

So, it's important for leaders, first, to recognize that they are vulnerable to these different sorts of followers in this day and age. And second, in a more positive sense, they need to be able to sell their product—the goals they wish to achieve—with their audience in mind.

And it is also very important for the followers. Most of us, even those who are quite near the top of the hierarchy, are followers because we have superiors. Hospitals, in particular, certainly in the United States, are traditionally very hierarchical. It is high time that followers and subordinates had a fuller understanding of what they are able to accomplish. There are certain tactics and strategies, ways of being heard, that followers in hierarchical organizations, who historically have generally been rather timorous, can now start to understand.

妳剛才談到「死硬派」，那個對理念等死心塌地的極端案例。妳說的，是組織裡潛在的告密者嗎？

是的，這些告密者本身也是一種極端。我們時常將這些告密者視為一種英雄，他們膽子很大，隨時都敢挺身而出，大聲張揚壞消息，就連因此惹禍遭殃也在所不惜。不過，我們要很小心，因為這些告密者一般下場都不會很好。

首先，他們說的話時常沒被聽見，所以他們的努力往往白費。其次，他們常常為自己的所作所為付出極高的代價。沒錯，他們扮演的是一種重要角色；我無意貶低這種角色的重要性。但我建議追隨者在扮演這種角色時，要格外謹慎小心，在可能毫無支援的情況下站出來以前，不妨先試試其他策略。不過，你說的沒錯，告密

者可能是「死硬派」。他們最後往往付出最高代價，遭
到組織開除。組織對他們也可能採取一些比較輕微的處
罰，如將他們邊緣化、降級，或者刻意忽視他們。所以
說，這是一種風險很高的追隨者類型，我認為在決定扮
演這種孤軍奮戰的角色以前，應該先試試其他辦法，這
是我的建議。

We talked earlier about the extreme case of the diehard—someone who believes passionately in the cause. So are you talking potentially about a whistle-blower?

Yes, the whistle-blower is an extreme case in point. We tend to think of the whistle-blower as being a heroic figure, as being bold and daring and being ready, willing, and able to go out and trumpet the bad news, even at the sacrifice of his or her own well-being. One has to be really careful, however, because whistle-blowers do not in general fare well.

First of all, they are often not heard, so their efforts are wasted. Second, they typically pay a high price for what they do. Now, it's an important role to play; I don't diminish it. But I suggest that people undertake that role with caution, and explore other strategies before going out on their own and out on a limb. But, yes, a whistle-blower can be a diehard, as you suggest. They often pay the ultimate organizational price of being fired. Penalties of lesser degrees would be simply being marginalized, demoted, or overlooked. So, it's a high-risk strategy, and I think there are other strategies that should be followed before one decides to play the lone ranger role. That would be my advice.

———

那麼，該用什麼策略呢？

第一步是在知識上確實掌握 21 世紀出現的變化，這類例子在美國很多。舉例來說，我在哈佛大學當教授的時候，當時的校長勞倫斯・桑默斯（Lawrence Summers）就遭到手下教職人員趕下台。或許你還記得，桑默斯當時說了一些關於女性能力的話引發爭議。這便是領導人遭到追隨者罷黜的一個好例子。

所以，注意追隨者這個議題，不僅合乎潮流，事實上也是明智之舉。今天，身為領導者的人，必須比過去更加注意部屬。身為部屬的人，若能更進一步了解自己可能扮演的角色，對自己也絕對有利。這不一定得是一種對抗性的角色，如果你認為自己的領導人很好，積極支持這位領導人對你就會很有利。如果你認為自己與老闆或管理階層有點問題，身為追隨者的你，可以找一些與你志同道合的同事，組成團體或聯盟，想出一些可能圓滑

解決問題的抗議之道。

所以強調追隨，目的就在於讓我們對領導人與追隨者之間的互動關係，有更成熟、更多的了解。這個趨勢今天在全球各地都出現，例如在中國，領導就受到僅僅 10 年以前已經難以想像的挑戰。而無論在英國或美國，醫療體系的傳統階級架構，自然也不能置身這波挑戰之外。

What sort of strategies?

The first step is really coming to grips with an intellectual understanding of the changes that have taken place in the twenty-first century. We have ample examples in the United States. For example, I was a professor at Harvard University when the university's then president, Lawrence Summers, was upended by his faculty. You may recall that he made some controversial comments about women's abilities. Here is a good example of a leader who was undone by his followers.

So, increasingly it is not only fashionable, but actually smart to pay attention to followers. As a leader today, you have to pay greater attention to those who are your subordinates. And as far as the subordinates go, it is absolutely in their interest to have a greater understanding of the role they can potentially play. This does not necessarily have to be an adversarial role. If you think you have a good leader, it is very much in your interest to be actively supportive of that leader. If you think you have some problems with your leadership or management, then as a follower you can connect to colleagues, people who feel the same way, forming groups, forming alliances, and figuring out how some protests might work diplomatically.

So the emphasis on followership is bringing a far greater level of sophistication and understanding of the leader-follower dynamic. And this is happening the world over, including in places like China, where the leadership is being challenged in a way that was unimaginable even a decade ago. And this certainly does not exclude traditional hierarchies in the medical system, whether in the United Kingdom or the United States.

——

妳先前寫了一本有關壞領導人的書，我們很好奇這本書與追隨者有什麼關係。好的追隨是壞領導的解藥嗎？

你這個問題問得太對了，讀過我那本有關壞領導人的書的人都知道，書中四大章每章都分為三節，一節談背景環境，一節談領導人，第三節談追隨者。寫那本書，讓我無論在實際上或在知識上，都了解追隨者的重要性。

這世上沒有壞領導單獨存在於壞追隨之外。有壞的領導人，就一定有支持、援助這類壞領導人的壞追隨者。當我在那本書提出這個問題時，同時也提出為什麼追隨者要追隨這些壞領導人的問題。我們人類已經能夠投入大把金錢對抗身體上的疾病，但我們一直未能對壞領導人這個社會疾病發動攻勢。在最惡劣的情況出現時，壞領導能夠致人死命。與你的問題暗示的完全一樣，我的答案是：除非有好的追隨者，想讓壞領導人停下來，甚至只是緩下腳步，都不可能辦到。事情就這麼簡單。

We are curious to know how followership is linked to your earlier book about bad leaders. Is good followership the antidote to bad leadership?

You're absolutely right to ask the question, because if one knows my book on bad leadership, one knows that each of the four chapters was divided into three sections, one devoted to the context, one devoted to the leader, and the third devoted to the followers. Writing that book made me understand actually as well as intellectually the importance of the follower.

There is no such thing as bad leadership without bad followership. If you have a bad leader, there must be bad followers who go along to assist and support that leader. When

I raised the question in that book, I raised the question of why followers go along with these leaders. The human race has been able to attack physical disease and throw money at it, but we have been unable to attack the social disease of bad leadership. In its worst form, bad leadership has lethal implications. My answer is exactly as your question suggests: you cannot stop or even slow bad leaders unless you have good followers. It's that simple.

——

▶ 合作創造偉業

　　學者們對合作真實本質的新探討，也讓我們對追隨愈來愈重視。特別是歐洲工商管理學院組織行為學教授，同時也是考拉領導與學習講座教授荷蜜妮亞·伊巴拉（Herminia Ibarra），更以她對合作式領導的研究成果，敲響命令控制式領導的喪鐘。伊巴拉早期的工作，以職涯與人才管理為重點，這是她另一個主要的研究領域。她在《工作認同：再造職涯的非傳統策略》（*Working Identity: Unconventional Strategies for Reinventing Your Career*）這本書中，詳細說明如何在職場中自我再造。近年來，她的研究重心轉向領導，特別是人脈與合作式領導。

伊巴拉與歐洲工商管理學院社會創意中心（Social Innovation Centre）兼任教授暨資深研究員馬克‧杭特（Mark Hunter）合作，研究領導人如何建立人脈。他們發現，有 3 種值得注意的基本人脈型態：作業型人脈（領導人需要這類人脈來順利推動日常作業）、私交型人脈（在組織外由志同道合之士所組成的人脈圈，有助於個人晉升），以及策略型人脈（在人脈網建立人控制範圍以外的人，可以協助他／她達到關鍵性組織目標）。伊巴拉對領導人需要什麼技巧，才能運用這些人脈的問題，也進行了探討。

她與歐洲工商管理學院企業學教授莫頓‧韓森（Morten Hansen）合作，觀察合作型領導人的組成要素。伊巴拉說：「沒有一家公司能夠自備一切所需資源，所以我們必須跨界工作。合作型領導人的精義就在這裡，它講究的就是動員，鼓勵員工跨越疆界取得豐碩成果。但什麼樣的領導，才能讓組織找出有意義的合作機會，將最優秀的人才引進組織，達成最有效的成果？」

伊巴拉認為，領導人如果想成為好的合作型領導人，就必須注重幾個領域。首先，他們需要建立良好的人際網絡，讓他們透過聯繫，以合作的方式增加價值。他們也應

該在廣大周邊上，接觸各式各樣的人才。伊巴拉說，所謂廣大「周邊，可能指其他地理區、其他國籍或其他世代，為的就是讓其他人加入討論，要有性別上的多樣性，這可能包含了許多面向」。

第二步則是為合作程序創造有利的條件，這階段的工作包括根除一切可能有礙合作的政治角逐與勢力範圍傾軋。伊巴拉說：「你必須站在最高點帶頭示範，如果你不能站在最高點展現你的合作潛能，合作不可能出現。」最後，領導人必須展現強勢的一面，並不是每件事都需要合作，一味講求共識只會毀了合作。合作型領導人不會一味講求共識，他們知道該在什麼時候退步、什麼時候採取行動，懂得適時推動合作程序，為組織增加價值。

為了有效掌握合作型領導，領導人必須否定有關領導的一些一般性成見。以情境領導與在某些狀況下的命令控制領導為例，伊巴拉說：「有關情境領導的理念已經根深柢固。許多人相信，時機好的時候，他們可以做一切好事，可以放手永遠合作。一旦時機不好，抓緊控制的時候便到了，必須善用命令與控制才行。但事實並非如此，在時機漸趨困難的時候，我們更需要好點子，更需要進一步向外求援。許多人仍認為時機不同，應該採取的做法也不

同,時局艱難的時候,就必須運用命令與控制。我認為真正的障礙,就是這種意識。」

▶ 與一流策略大師對談

蓋瑞‧哈默爾(Gary Hamel)在接受我們訪談時,重申了伊巴拉有關合作型領導的觀點。下列是我們與哈默爾的訪談要點。

依你之見,身為領導人,最重要的事是什麼?

把所有商管書依主題分類、堆成幾堆,有關領導書那堆會比談策略、談變革或其他主題的書堆大很多。深入公司內部觀察,為了培養具有某一套能力的領導人,我們花了許多時間。我們期待的領導人既大膽又謹慎,既強壯又富有同理心,既有遠見又重實際。但事實的真相是,這樣的領導人少之又少。

What do you see as the most important thing for a leader?

Pile up all the different business books by subject area, and the books on leadership make a bigger stack than those on strategy, change, or anything else. Look inside companies, and we spend a lot of time trying to produce leaders who have a certain set of capabilities, people who are bold yet prudent, strong and yet empathetic, visionary yet practical. But the fact of the matter is, there are very few of those people out there.

———

既然這樣的領導人很少，組織應該怎麼做？

我們需要問的一個問題是：我們應該注意的，是尋找、培養傑出領導人，還是應該想辦法建立好的組織，就算領導層能力平庸，組織仍能表現亮眼？

我認為，後者是比較可行的辦法。無論你用什麼標準進行評估，過去一世紀以來，民主體制的表現總是超越獨裁系統。

在觀察這些資料時，你會因民主展現的彈性與調適性而動容。在民主體制，權力由下而上，責任由上而下。在公司的情況，則通常正好相反。所以，我的意思並不是說，我們不應該努力改善我們的領導技巧與能力。我是說，賦予少數高階主管一大堆策略與決策權威，將他們視為超人的觀念，到頭來總是會徹底破產、完全行不通的。

So if there are not a lot of leaders like this, what should organizations do?

One of the questions we have to ask is, is the problem finding or growing these extraordinary leaders? Or is it building organizations that can thrive even when they have fairly mediocre leadership?

I think it's much more the latter. If you look at any measure, democracies have outperformed totalitarian systems over the last 100 years.

When you look at the data, the thing that strikes you is that democracies are resilient and adaptable. In a democracy, power flows up and accountability flows down. In companies, it tends to be exactly the opposite. So it's not that we shouldn't strive to improve our leadership skills and capabilities, it's just that at the end of the day, I think that the notion that we're going to invest a lot of authority over strategy and direction in a small group of people at the top who are somehow superhuman is an entirely bankrupt notion.

———

這麼說，我們不能只是依賴偉大的領導人，帶領組織度過商業困境？

我們已經知道，無論任何社會系統都有一個共同性，就是權力愈是集中的系統，調適能力愈差。許多公司為重振昔日雄風而掙扎不已，但它們面對的問題其實不是不景氣，而是商業環境本質上的轉型。它們的商業模式已經過時了，這類公司多得不勝枚舉。

為什麼會出現這種狀況？10 次裡面有 9 次，原因就在於這些公司的領導人，仍然守著他們那套已經貶值的智慧資本不肯放手。他們周圍的世界不斷在變，但他們仍然堅持那套無論對客戶、科技或商業模式來說，都是早已過時的信念。他們仍然大權在握，仍然高高在上。既不肯調適、不願改變，組織當然也欲振乏力了。

So there is no use just relying on great leaders to steer organizations through a difficult business environment?

One of the things we know about any social system is that the more highly you concentrate power in it, the less adaptable it is. There is a long list of companies that are struggling to regain their mojo where the problem isn't the recession, the problem is a fundamental shift in the business environment. They have a business model that's out of date.

Why did that happen? Nine times out of ten, it happened

because the people at the top were unwilling to write off their depreciating intellectual capital. The world was changing around them, but they hung on to out-of-date beliefs about the customer, the technology, and the business model. They still had the power and the authority, and so the organization's capacity to change was held hostage to their personal willingness to adapt and change.

──

如果刺激改革所需的動力，不能來自組織領導人，它應該從何而至？

我們需要大幅減少對領導的關注，需要更加重視如何在日常工作上，運用組織集體智慧與活力的問題。我們應該怎麼整合這種智慧與活力，才能知道下一步應該怎麼走、知道新機會在哪裡？有人認為，我們應該靠一、兩個最高領導人，來替我們做決策、替我們訂定未來願景。我認為，在今天這種複雜而迅速變化的世界，這樣的想法已經行不通了。

我常發現一個令人深思的問題：儘管許多執行長也經常談到公司需要改革，並談到改革為什麼那麼重要，但據我所知，幾乎沒有一家公司已經為每位員工培訓，讓

他們成為內部活躍分子。事實上，我們對這種構想幾乎是望而生畏。讓每位員工活躍，並不意指一種無政府狀態，它指的是活躍。在民主國家，改革總是由平民百姓而起，平民百姓組成新政黨，或展開一項運動以拯救環境等。但在組織內部，我們仍然非常欠缺這種鼓勵一般員工挺身而前、建立聯盟、推動觀點、訂定政策的能力。在我們的組織，活躍分子只能無助地呆坐在那裡，因為組織告訴他們，策略是最高領導人的事，與他們無關。

這種情況會成為一種自我實現的預言。如果不對任何策略負責，若想改變，你就只能等到下一位領導人上台以後再說。我們必須改變這種現狀，因為如果不能改變這種現狀，在錯失環境中不斷變化的重大事物之後，我們的組織必將經歷一波波持續不斷、痙攣似的改變週期。

If not from leaders, then where is the impetus for the kind of change that is required from organizations going to come from?

We need a lot less attention on leadership and a lot more on how do you, on a day-by-day basis, exploit the collective intelligence and energy of the whole organization. How do you aggregate that intelligence and energy in a way that reveals where we should go next, the new opportunities? But as for the idea that one or two people at the top are going to be the primary decision makers and visionaries, I think that's just unsustainable in a world that is as complex and is changing as fast as the one we have right now.

I've often found it interesting that, while a lot of CEOs will talk about the need for change and how it's so important, I know of virtually no company that has trained every employee to be an internal activist. In fact, we're almost afraid of that kind of idea. This doesn't mean an anarchist, it means an activist. In democracy, change starts with ordinary people who form a new political party or start a campaign to save the environment or whatever. Yet inside organizations, we really haven't harnessed that ability of frontline people to build a coalition, advance a point of view, and start to shape policy. They're sitting there helpless. They've been taught that strategy starts at the top.

That turns out to be a self-fulfilling prophecy. If you don't take any responsibility for strategy, you have to wait until the next leader comes in. We have to change this, or our organizations are simply going to go through continuous cycles of convulsive change after they have missed critical things that are changing in the environment.

———

第 7 章

領導人
面對世界

華頓商學院「領導力學程」
創辦總監
與談人｜ **史都華・費德曼**

我們遇到的所有成功領導人，都有一種正面、樂觀的世界觀。他們並非不切實際，只是在觀察事情的時候，總喜歡從正面著眼，看到的會是裝了半杯水的杯子，不是只剩半杯水的杯子。

近年，由於「正向心理學」（positive psychology）的崛起，「正向」現在已成了愈來愈時髦的詞彙，愈來愈多人開始將它運用於領導。西班牙 IE 商學院（IE Business School）的李・紐曼（Lee Newman），就是正向領導的先驅之一。紐曼以取得行為優勢為目的，提出一種新領導做法，主張「建立一個由個人與團隊組成，在每個層面上都能想得更好、表現得更好的組織，從而取得優勢」。

紐曼曾經創辦過一家科技公司，也曾擔任麥肯錫（McKinsey & Co.）顧問，他認為公司已經不再能保住傳統意義的永續競爭優勢，但或許能保住行為優勢（behavioral advantage）。紐曼認為，只要運用行為經濟學與正向心理學最新的研究與思想成果，就能改善個人與組織的表現，取得這種行為優勢。

根據紐曼的見解，正向領導有 3 大要件。第一項要件是思考訓練，這能協助領導人了解他們的決策思考過程，

讓他們思考得更加周詳。第二項要件以強化員工的長處，
而不以改善他們的短處為重點。紐曼說，公司應該「找出
員工與團隊的長處所在，然後根據這些長處，為他們設計
工作。這是一種雙贏的做法，不但對員工的福祉比較好，
對組織的利益也比較好」。

正向領導的第三項要件是，領導人需要常保專業體質
的健全，確定自己與部屬在日常工作中不斷運用所學。紐
曼指出：「之所以稱為『正向』，是因為它講究的，就是
讓表現已經不錯的專業人士表現得更好，沿著上升曲線邁
向優異。」能夠滿足這 3 大要件的正面領導，「是能夠在
現代職場中協助公司，幫公司達到永續性優異成果的新途
徑」。

▶ 領導，是為了帶領眾人過更好的生活

賓州大學華頓商學院管理學客座教授史都華・費德
曼（Stewart Friedman），畢生與正向領導結下不解之緣。
他是組織心理學家，也是「華頓領導力學程」（Wharton
Leadership Program）的創辦總監，以及「華頓工作／生活整
合計劃」（Wharton's Work/Life Integration Project）的創辦總

監。此外，費德曼也是「全面領導」（total leadership）概念的創始人，著有《全面領導：做更好的領導人，過更好的人生》（*Total Leadership: Be a Better Leader, Have a Richer Life*）一書。

> **領導學現在已經成為一種重工業，但在你投入這門學術領域的時候，它還只是一個看起來幾乎不值得認真研究的東西。你對領導有什麼看法？**

30 年前，我在密西根大學的論文研究主題，就是大公司如何培訓、選用主管。之後，我進一步研究主管發展、接班計劃、執行長接班，以及領導力發展系統。1991年，華頓大舉翻新課程，我便在華頓開了「領導力學程」。當時，我們的課有一項關鍵要件，就是讓企管碩士班全部 8 百多位一年級生，每個人都參與團隊，在一種團隊環境中就領導議題提供回饋，從而取得實務經驗。我們創造了學習團隊，在那個時候，學習團隊在企管碩士班課程中是一項創舉。

Leadership has become a heavy industry, but when you started out, it was barely recognized as being worthy of serious study. How did you come to look at leadership?

My dissertation research 30 years ago at the University of Michigan was about how executives in large companies are prepared and selected for their positions. That led to my work on executive development, succession planning, CEO succession, and leadership development systems. In 1991 I started the Wharton Leadership Program. That was when Wharton did a major revamping of our whole curriculum. One of the critical elements of that work was to ensure that our students—more than 800 first-year MBAs—had real-world experience working together in teams and providing feedback on leading in a team environment. We created learning teams, which at that time were innovative in an MBA curriculum.

———

直到 1980 年代，領導才真正起飛。

當然，早自柏拉圖起，世人就在思考如何培養未來領導人的議題，還為這個議題爭執不已。領導不是一個新議題，不過直到 1980 年代，它才在現代商界起飛。

儘管 1960 年代與 1970 年代，也出現許多討論領導的著作，但是劃時代性的作品是湯姆‧彼得斯（Tom Peters）與羅伯‧華特曼（Robert Waterman）的《追求卓越》（*In Search of Excellence*）。他們是首先端出文化與領導完整議

題，探討領導在組織表現所扮演角色的第一人。

人性潛能運動是又一重要先驅。在這項運動引領下，我們開始專注於生活品質，開始考慮完整人生的問題，開始將領導視為一種人人都可以參與的事——它不只是組織階級中一項主管角色，而是一種能為更大的社會使命或運動貢獻的個人能力。

It was really in the 1980s that leadership took off.

Of course, people have been thinking about and struggling with the issue of how you develop future leaders since Plato. This is not a new issue. But in the 1980s, it took off in the modern business world.

Although there had been an emerging literature bubbling up through the 1960s and 1970s, the watershed book was Tom Peters and Robert Waterman's *In Search of Excellence*. They were the first to shine a bright light on the whole issue of culture and leadership, and the role of leadership in organizational performance.

The whole human potential movement was another important precursor to a focus on quality of life, on taking the whole person into account and seeing leadership as something that was available to everyone—not as just an executive role in a hierarchy but as a person's capacity to contribute to some larger social mission or cause.

———

在你的研究工作中，最有意思的部分是工作／生活整合與領導之間的關聯。

我一直認為，想解決工作／生活這個兩難議題，就得靠領導，而且領導的核心其實就是全人的概念。

我認為，我們可以在職場與其他地方提升我們的領導能力、表現與成果，以一種智慧的方式，將人生各個不同部分整合在一起，讓我們的人生更加美滿。我希望，我已經證明這個構想確實可行。這個構想的基本概念是，領導就像任何表演藝術或運動一樣，就算不能傳授，仍是一種可以學習、操作與發展的能力，每個人都可以將領導人的角色扮演得更好。

What's really interesting in your work is the link between work/ life integration and leadership.

I have pursued the idea that the solution to the work/life dilemma is leadership and that the heart of leadership is really the whole person.

I have, I hope, demonstrated and brought to life the idea that you can advance your leadership capacity, performance, and results at work and elsewhere by bringing together the different parts of your life, integrating them in an intelligent way that works for you. Behind this is the notion that each person can emerge as more of a leader than he or she currently is, and that leadership can be learned, practiced, and developed, like any performing art or sport, even if it cannot be taught.

———

工作／生活的辯論往往呈現兩極化，成為一種性別議題。你的看法非常不一樣。

儘管不幸的是，工作／生活的兩難，時常遭人加上女性議題的標籤，但它絕不是女性議題，而是一種社會、人性與經濟議題。我之所以提出「全面領導」這個構想，目的就是要刻意以顛覆手段建立一種語言，讓男性都能運用這些原則與方法來進行全面領導，進而整合、提升他們的生活品質。

所以說，它談的是領導、表現與人生中每一部分可衡量的成果，包含工作、家庭、社群與私我（身心靈），也就是我所謂的「四面皆贏」。無論男女，它應該都是一種能讓人興奮以求的東西，或許更重要的是，它是一種

能讓人放心追求的東西。

The work/life debate is often polarized as a gender issue. You take a very different perspective.

While it has often and unfortunately been characterized thus, the work/life dilemma is certainly not a women's issue—it's a social, human, and economic issue. What I was intentionally, and subversively, trying to accomplish with the idea of *total leadership* was to create a language that enabled men to access these principles and methods and use them to better integrate their lives, and for it to be legitimate for them to do so.

So it's about leadership, performance, and measurable results in all parts of life: work, home, community, and the private self (mind, body, and spirit)—what I call "four-way wins." That's something that a man, as well as a woman, can get excited about pursuing and, perhaps more important, can feel comfortable pursuing.

———

這許多構想的源起，可以回溯到 1999 年你在福特公司工作的時候。

是的，那段時間讓我幾乎整個思想都轉了型。我在那家

了不起的公司，當了兩年半資深主管，我的世界也因這兩年半的工作經驗完全翻轉。現在回想起來，那幾乎已是 15 年前的往事。回憶起那段經驗，我要說的第一件事，就是面對在一個大型社會系統中完成任務的艱鉅挑戰，我覺得自己卑微之至。

在離開福特時，對那些投入畢生心血領導組織、想方設法完成任務的人，我有一種與過去截然不同的敬佩與仰慕之情。他們的工作，比外人眼中看來要難得太多了。

The genesis of many of these ideas can be traced back to your time working with Ford, starting in 1999.

Yes, that absolutely transformed my thinking about just about everything. My world was completely turned around by my experience as a senior executive in that incredible company for that two-and-a-half-year period. The first thing that I would say about that experience, reflecting back now almost 15 years, is how humbled I was by the challenge of trying to get anything done in a large social system.

I came away with a profoundly different kind of respect and admiration for people who spend their adult lives leading organizations and trying to get important things done in them. It's so much harder than it looks from the outside.

——

你當時負責福特編制 50 人、作業預算 2,500 萬美元的領導發展中心。後來怎麼樣？

傑克‧納瑟（Jac Nasser）在出任福特汽車執行長時，想改變這家公司的文化，也想改變員工的心態，要他們把更多注意力擺在消費者端，而不是製造端。他要公司每一位員工都能勇於面向外界、勇於探討環境，還要求員工自視為自己人生各種不同部分的領導人。

在這份工作的聘用面談中，我告訴傑克，如果他聘我主持這個中心，我要使領導發展成為一種全人的概念，而不是只談公務，其他什麼都不管。傑克聽了很高興說：「太好了，我愛這一套！」我就這樣開始上班了。

直到今天，還是有許多人認為這些事，不是公司應該擔心的問題。不過，我們花了很長時間、投入許多心血，看到今天的世界，就領導發展的傳承與價值而言，竟與過去有這麼大差異，實在令人稱奇。今天，「全面領導」無疑已是一個大家都認同的理念。看到許多男性也能比過去更自在地討論工作／生活的難題，也能從全人的角度來觀察領導，讓人很開心。

You were running the company's Leadership Development Center, a 50-person, $25 million operation. How did that come about?

When Jac Nasser became CEO of Ford, he was looking to transform the culture of the company and to change the mindset of employees to focus more on the consumer and less on the manufacturing side. He wanted everyone in the company to be outward-facing, to really take in the environment, and to see themselves as leaders in all the different parts of their lives.

Interviewing for the job, I told Jac that what I was going to do here, if he were to hire me, was to make leadership development be about the whole person, and not just about business. And when he said, "Great, I love it!" that sealed the deal for me.

It's still contrary to what a lot of people think companies should be worrying about. But we have come a long way. It's amazing to reflect on how different the world is today with respect to the legitimacy and value of leadership development. It's definitely an idea that is now fully embraced. It's also heartening to see how men are now much more comfortable talking about work/life dilemmas and seeing leadership from the point of view of the whole person.

———

但大家仍在抱怨，說找不到領導人。

是的，這是事實。全面領導的主要目標，就在於探討我
們對領導能力發展的認知，讓我們有信心、有能力在人
生中進行有意義且可以持續的改革。但在這個過程上，
我們還有很長遠的路要走。由於面對這個議題的態度不
夠認真，我們浪費的努力與人才仍然太多。不過，這不
表示我們沒有多大進展。

And yet people still complain that there's a shortage of leaders.

Yes, that's true. We have a long way to go in terms of how we
think about the centrality of developing leadership capacity and
giving people a sense of confidence and competence in creating
meaningful and sustainable change in their lives—a primary
goal of the *total leadership* approach. There's still so much effort
and talent being wasted as a result of our not taking this issue
seriously enough. But that's not to say that we haven't made a lot
of progress.

——

你目前的研究對象是什麼？

許多人常問我一個問題：「全人的領導觀在理論上很不
錯，但史都，在真實世界中，除非你專心致志投入一件

事，否則你什麼事也做不好，不是嗎？」我一直要求學生與其他人寫小傳，討論能夠實踐全面領導原則、讓他們仰慕的人——那些能夠了解對他們而言，什麼是「真」的人（由於認清哪些事對自己重要，而能真誠以對）；那些能夠了解對他們而言，什麼是「完整」的人（由於能尊重全人概念，而能正直以對）；以及那些能夠了解對他們而言，什麼是「創新」的人（能以創意行動、不斷實驗完成工作的方法）。這些人之所以能在世上成就豐功偉業，並非是因為他們犧牲了人生的其他部分，而是因為他們懂得如何照顧家庭、社群與私我，事業也因此受益。

我們常犯的一個迷思就是，想當一位成功的領導人，你就必須放棄人生的其他部分。當然，在若干程度上，做一些犧牲，總是在所難免。但這世上也有許多人能在人生的不同部分找出共同價值，從家庭、社群、情緒與精神生活中，取得力量、智慧與支持，從而在專業生活中開創佳績。這樣的例子很多；事實上，我的《偉大領導人，美好的人生》（*Great Leaders, Good Lives*）這本書，講的就是這些故事。

全面領導的要旨，是以一種有系統、有紀律的做法，了解什麼事對你和身旁的人最重要，然後進行實驗，創

造我所謂的「四面皆贏」。我們已經找出一套可以實踐「真」、「完整」與「創新」3 項基本原則的技巧。《偉大領導人，美好的人生》一書所討論的，就是 6 位享有美滿人生的偉大領導人，書中還會分析他們運用的技巧，讓讀者知道任何人都可以運用這些技巧。

What is your current research looking at?

One question I get asked a lot is this: "A whole person view of leadership sounds good in theory, Stew, but in the real world, you can't really do anything great unless you're completely and fully dedicated to it, right?" I have been asking students and others to write short biographies of people whom they admire and who exemplify the principles of *total leadership*: who have figured out for themselves what it means for them to be real (to act with authenticity by clarifying what matters to them), to be whole (to act with integrity by respecting the whole person), and to be innovative (to act with creativity by continually experimenting with how things get done). These are people who have achieved greatness in the world not in spite of their commitments to other parts of their lives, but because of how their work benefits from their investments in family, community, and the private self.

The mythology is that you have to write off the rest of your life in order to have success as a leader. Of course, you always have to make sacrifices at some point. But there are many examples of people who have drawn power, wisdom, and support from their families, from their communities, and from their emotional and spiritual lives in order to achieve great things in their professional lives by finding mutual value among the different parts of life. Indeed, this is the story I tell in my book *Great Leaders, Good Lives*.

The *total leadership* message is to have a systematic and disciplined approach to focusing on what matters most to you and to the people around you, and then experimenting with ways of creating what I call four-way wins. We have identified a set of skills that bring to life the three main principles of being real, being whole, and being innovative. *Great Leaders, Good Lives* will feature six great leaders who've lived good lives and an analysis of the skills they've used to do so. It will show how anyone can practice those skills.

———

你目前仍是「華頓工作／生活整合計劃」的領導人。

「華頓工作／生活整合計劃」創辦於 1991 年，有兩個主要任務。一個任務是將商界、學界與公共領域的思想

領導人聚在一起，讓他們找出如何在個人層面、執行層面、組織與社會層面，整合人生不同部分的最佳做法，並且共享這些做法。

這個計劃的另一項任務，是對華頓學生與校友的人生與職涯進行研究。這些研究的第一件產品，是 2000 年出版的《工作與家庭——是敵是友？》（*Work and Family—Allies or Enemies?*）。我們在 1992 年，蒐集了對畢業班學生的深入調查資料，20 年後，我們對 2012 年畢業班學生進行同樣的調查。然後，我們回頭找上 1992 年的這些校友，向他們提出有關生活、職涯、成就與未來抱負的進一步問題。我們問他們如何因應工作／人生挑戰，問他們對配偶雙方都在工作的雙職涯關係的看法，並了解他們的工作如何變化，以及科技對他們有什麼影響等。

我們對兩班學生提出一個共同問題：你計劃生養孩子嗎？在 1992 年，有 79% 的男女學生說「是」；到了 2012年，只有 42% 的學生說「是」。我們出版了另一本書，叫做《嬰兒荒》（*Baby Bust*），觀察 20 年來工作與家庭生活出現的巨大變化。我在書中描述事情為什麼轉變、如何轉變，並指出轉變的原因與方式，因性別不同而大不相同。舉例來說，計劃要生養孩子的男性比過去少，就與計劃要生養孩子的女性比過去少的原因不一樣。

好消息是，男性與女性對工作與家庭生活的態度與價值觀愈來愈接近。全面領導計劃提出的理念與方法，能夠幫助他們創造一種新社會秩序，讓他們透過學習而發現，他們能用來推動改變的力量與辦法，其實比他們想像的多很多。一旦有了這種覺悟，他們就能更勇於參與對他們最重要的事，讓自己及身邊的人過得更有意義，從而對世界做出更大貢獻。

這一刻的歷史意義之所以這麼令人振奮，是因為我們正進一步提升人類解放的理念。在 1960 年代當我還年輕的時候，主張解放人類精神的運動，在我們的社會蔚為風潮。這場人性革命的第二階段，現在已經展開。你看，現在有這麼多探討家庭生活方式的實驗，有這麼多年輕人提出「什麼可能、什麼可以接受」的質疑。當然，數位革命的進展，也讓一切加快了腳步。在如何工作與貢獻的議題上，無論男女都會有多出許多的選項。

You are also still leading Wharton's Work/Life Integration Project.

The Work/Life Integration Project started in 1991 and had two primary missions. One was to bring together thought leaders in business, academia, and the public sector and to have them discover and share best practices in integrating the different parts of life at the individual level, the executive level, and the organization and societal level.

The other part of the project was research on the lives and careers of Wharton students and alumni. The first product of that was the 2000 book *Work and Family—Allies or Enemies?* In 1992 we gathered in-depth survey data from the graduating class, and 20 years later we administered the same survey to the graduates of the class of 2012. Then we went back to the class of 1992 and asked them further questions about their lives, careers, achievements, and future aspirations. And we asked them how they deal with work/life challenges, their views on dual-career relationships, how their work has evolved, how technology has affected them, and so on.

One of the questions we asked of both classes was, do you plan to have children? In 1992, 79 percent of men and women said yes, and in 2012, 42 percent said yes. Another book, called *Baby Bust*, looks at how the past two decades have seen enormous changes in both work and family life. I describe why and how things have changed, and in very different ways for men and for women. For example, the reasons why fewer men are planning to have children are not the same as the reasons why fewer women are planning to have children.

The good news is that there is a growing convergence between men and women in terms of their attitudes and values concerning both work and family life. And the ideas and methods in the *total leadership* program can help them create a new social order. That is, among the key takeaways from the *total leadership* program is that people learn to recognize that they have a lot more power and discretion to create change and to harness that power. This enables them to attend to the things that matter most to them and to make a greater contribution to the world in ways that work for them and for the people around them.

What's so exciting about this moment in history is that we're taking the idea of human liberation one step further. Earlier in my life, in the 1960s, there was an explosion of interest in liberating the human spirit. The next phase of our evolution is under way. There's so much experimentation now in terms of the ways that families live their lives, with many young people questioning what's possible and acceptable. And, of course, the advent of the digital revolution has accelerated everything. There are going to be a lot more options available to people, men and women, for how they choose to work and how they contribute.

——

你認為自己是領導人嗎？

我希望每個人都將自己視為領導人，這當然也適用於我自己。我對領導人做些什麼事的主要概念就是，領導人會把一些人結合在一起，鼓勵他們行動，讓他們完成有價值的目標。這些目標必須能增進他人福祉，才算有價值。這是我的工作任務，這是我無論在哪裡——無論面對華頓的學生，或在其他任何地方——都會全力鼓吹的理念。我們每個人，無論在工作場合、家裡、社群或獨處時，都需要做我們自己。在人生每一個重要領域，我們都應該展現完整的自我，讓這個世界更加美好。

Do you think of yourself as a leader?

I want everyone to think of him- or herself as a leader, and certainly that applies to me too. My main conception of what leaders do is that they bring people together, they mobilize people, to accomplish some valued goal. And for it to be valuable, the goal has to make things better for other people. So that's my mission through my work, and that's an idea that I'm trying my best to help cultivate wherever I go—with our students at Wharton and beyond. Each of us needs to be him- or herself in whatever sphere we're in—at work, at home, in the community, and alone with our private thoughts. We need to bring ourselves in full to all the important domains in our lives to make the world better.

———

你的作品有一種很濃郁的樂觀氣息。

我很高興你提出這一點，因為對我來說，這是身為領導人必須做的事：必須將今天殘酷的現實，轉換為一條充滿希望的途徑，讓世界在明天會更好一些。身為領導人，必須盡可能將現實看得清清楚楚，然後與其他人一起以開創精神，想出改善人類條件的辦法。

There is a powerful strain of optimism in your work.

Well, I'm glad you picked that up, because to me, it is the hallmark of what leaders have to do: to convert the harsh realities of today into a hopeful path to make the world a little better. It is about looking at reality as clearly as you can, and then creatively, and in concert with other people, trying to figure out ways to improve the human condition.

———

第 8 章

領導人實踐

《UP 學》作者
馬歇爾・葛史密斯
麻省理工學院訪問學者
與談人│**凱特・史威曼**

有意學習領導的人，可以取用的資源很多。事實上，正因為資源太多，最大的挑戰或許就是你得知道從何著手。本書提到的，都是在各自領域最受尊重的領導思想家，無論你列的是什麼領導導讀清單，少不了會出現他們撰寫的許多書籍與文章。

但當領導人或有志於領導的人，在看完數不清的領導書刊，書架也因為承載太多領導經典著作，而似乎就要垮下來的時候，下一步該如何做？在做完了長篇累牘的筆記、畫完各式各樣的重點、在腦海裡裝滿構思、寫完一張又一張的小抄之後，領導人總得臨陣上場，做一番領導才行。

大多數有關領導的書刊，甚至一些最理論性的書刊，都會為領導人（或者應該說「追隨者」還更切合）提出一些實用建議。不過，有幾位領導思想家特別專注於領導實踐，他們對領導資質、特性與特質或許著墨不多，但對領導功能面與實用面有很深入的看法。

▶ 從軍隊走向董事會

以英國桑德赫斯特皇家軍事學院（British Royal Military

Academy）資深講師約翰・艾戴爾（John Adair）為例，軍隊是偉大領導行動的典型範例，偉大領導人能征服國家，差勁的領導人會讓部屬遇害，甚至連自己也送了命。艾戴爾專注於領導理論在戰場上的實用，自然可以理解。艾戴爾認為，他在軍中接受的領導訓練可以改進。他在日後寫道：「我在國民兵服役期間接受的少尉軍官訓練，總共列了 32 項身為領導人必備的資質。」

艾戴爾設計了一套確實可行的領導人訓練做法，取代這洋洋灑灑的資質清單。他指出，領導的關鍵功能為計劃、主動、控制、支援、通訊與評估。他以一套「行動中心領導」模式而著稱，根據他的觀點，領導是透過行動鍛鍊養成的能力——必須置於情勢中，讓領導人可以親歷領導經驗。他的模式是 3 個相互重疊的領導責任領域：團隊、任務與個人。艾戴爾說，領導人有責任協助一個團體完成任務，有責任將一個團體打造為一支團隊，有責任培養並鼓勵團隊的個別成員。

根據艾戴爾的模式，領導人的角色有相當一部分就在於維持這 3 項要件之間的平衡。團隊的勢力如果過大，團隊會墮落腐化，淪為一個委員會。太注重任務，領導人最後會成為獨裁者。如果不能對個人寄予適度關注，團隊

會秩序蕩然。1970 年代，艾戴爾將他的模式從軍官班教室搬進公司董事會。團隊、任務與個人 3 個相互重疊領域的概念，也濃縮為由 3 個互鎖在一起的圈所構成的文氏圖（Venn diagram）。這個 3 圈圖印在卡片上，分發給無數管理員與經理。

在商業環境下，「行動中心領導」模式注重的，就是讓經理人從行政管理人員轉型為領導人。艾戴爾這種以功能建立領導的做法，廣獲商界運用。之所以如此，部分是因為他非常支持公司內部的領導訓練，同時也因為他與產界淵源很深。艾戴爾認為，領導技巧必須實用，必須幾乎任何人都可以學得會。他對未來領導趨勢的預期，顯然比他那一代人先進得多。

艾戴爾說：「要在全球市場競爭、成長，公司必須注重創意與創新。要達到這個目標，公司需要以人為中心的領導人，而不是那種權威、跋扈的舊式經理人。問題是，現在有太多經理人把自己當成控制人、分配人或會計師。」

▶ 引爆點式領導：大刀闊斧的 NYPD 局長

艾戴爾運用他的軍事領導經驗，造就了更多有效的領導人。歐洲工商管理學院教授金偉燦與蘭妮‧莫伯尼（Renée Mauborgne），則從公共服務領域來觀察領導，領導力在這個範疇同樣具有關鍵地位。

2003 年 4 月，金偉燦與莫伯尼以「引爆點式領導」為題，在《哈佛商業評論》發表論文，並以紐約市警局長布拉登為所謂「引爆點式領導」的範例。1990 年代中期，紐約市犯罪事件爆增，紐約市警局面對薪酬低、工作環境危險、工時也過長的種種壓力，軍心潰散、萎靡不振。紐約市警局既欠缺足夠資源，又苦於組織政治傾軋。

根據金偉燦與莫伯尼的分析，布拉登的領導以 4 項要件為基礎：認識（溝通，確定管理人了解問題），政治（平息內部紛爭，孤立外來對手），資源（集中資源對付有問題的地區），以及鼓勵（鼓舞組織內部各階層士氣）。布拉登用了一些戰術，來落實策略、排除障礙。他迫使資深管理階層面對面處理問題，如統計數字顯示，紐約市地鐵是一項安全的交通工具，但許多實際搭乘地鐵的市民卻說搭乘地鐵不安全。布拉登便迫使各分局局長與中

層管理人員定期搭乘地鐵，讓他們親身接觸地鐵現況：幫派、強行乞討、逃票，以及其他各式各樣的犯行。

引爆點式領導人意圖改革，受到強大既得利益的阻撓自屬難免。以組織政治為例，有 3 種人物能造成不成比例的效應：天使（angel，因策略性移轉而獲利最大者）、魔鬼（devil，因策略性移轉而受害最大者），以及軍師（consigliere，深受敬重的圈內人，能指引引爆點式領導人走出政治地雷區）。

對引爆點式領導人而言，行動計劃的第一步，就是找一位深受敬重的軍師。其次是建立聯盟，爭取廣泛支持以孤立魔鬼，並且預估可能出現的反對意見，設計反擊對策。金偉燦與莫伯尼認為，引爆點式領導人特別重視對「鼓勵」要件有超大影響力的 3 種人與物：頭目、魚缸式管理與自動化。頭目就是組織內具有關鍵性影響力的人物，布拉登把注意力集中在 76 個警區的幾個頭目身上。他知道，透過這幾位頭目，他可以觸及 2 百到 4 百名高級警官，以及 3 萬 5 千名第一線警察。

接下來，透過透明化、包容與公平賞罰的過程，布拉登開始宣揚頭目們幹下的好事與壞事。這項過程幾乎立即

創立一種爭取表現的文化，因為頭目們最好面子，最怕在人前失了光彩。最後，引爆點式領導人會設法讓任務更容易管理。為了達到這個目的，布拉登將這項任務視為一種「一條街接一條街、一個警區接一個警區」的挑戰。採用這種觀點以後，在紐約這麼大的大城市壓制犯罪，似乎也沒那麼龐雜、艱鉅得讓人不敢面對了。

在應付資源挑戰時，布拉登把資源集中在幾個熱點上。所謂熱點，便是指需要的資源不多，但可以造成重大成果的活動重點。舉例來說，地鐵犯罪集中在少數幾條線與幾個車站，但在布拉登上任時，大多數地鐵線與車站配置的警力都一樣。而布拉登的解決辦法就是調集警力，集中部署在熱點上。另一方面，也要設法整頓耗用許多資源，但成果相對較少的活動。例如，過去在法庭審理嫌犯，即使只是最微不足道的犯行，警方也需要投入 16 個小時，這些時間主要花在將嫌犯從犯罪現場帶到法庭的過程上。於是，布拉登便推出「犯行巴士」（bust buses），直接在犯罪現場展開審理工作。

引爆點式領導果然有效，布拉登在僅僅兩年任期間，將紐約市重罪犯罪率減少 39％、謀殺罪減少 50％、盜竊罪減少 35％。他曾經在紐約市交警（New York Transit

Police）、波士頓大都會警察局（Boston Metropolitan
Police）、麻州灣區大眾運輸局（Massachusetts Bay Transit
Authority）、波士頓四警區與洛杉磯警察局（Los Angeles
Police Department）任職，他在這些組織都大舉推動改革，
而且都做得很成功。

▶ 學習當一個好老闆

　　超過 20 年以來，哈佛商學院企管系華勒斯・布雷特・
唐漢（Wallace Brett Donham）講座教授琳達・希爾（Linda
Hill），一直在研究升上領導職位的人，特別是那些非常有
潛力、初上管理階層的人才。她早期的研究以少數經理人
為對象，追蹤他們在管理職位上第一年的表現。這項研究
成果，成為她在 1992 年出版的《新手主管》（*Becoming a
Manager*）一書的一部分。

　　希爾發現，許多年來，經理人在第一個領導職位上
面對的挑戰已經愈來愈大。她說，認為這項職位轉型很難
的經理人之所以有這種感覺，一部分是因為他們對自己這
項新角色有若干誤解。新領導人認為新職位能為他們帶來
許多權威與權力，能夠隨心所欲地運用這些權威與權力。

事實上，他們往往發現，想做好領導人的工作，就得應付各種淵源與關係，而這些淵源與關係讓他們做起事來綁手綁腳、無法盡情發揮。能夠早點學會管理這些網絡般的關係，這些經理人也就愈能早點掌握新領導角色。

這些新領導人的另一個常見迷思是，以為擔任領導職位自然能為他們帶來權威。他們的直接部屬奉令行事，領導人要他們怎麼做，他們就會怎麼做。但新領導人很快就發現，事實並非如此。希爾說，新領導人需要展現擔任新角色的個性與能力，要表明自己有想做正確事情的意願。但這並不是說，新手主管必須有多高明的技巧，其實他們只要願意發問、傾聽部屬的建言就行了。此外，新領導人也需要展示，他們能運用在組織的關係發揮影響力。

這些新手主管常犯的一項錯誤是，走馬上任第一天，便立刻展示威風，讓團隊知道誰是老闆。透過正式權威控制部屬、讓部屬聽命，即使能做到，效果也無法持久。希爾說，與部屬共享權力與影響力，比發號施令要強得多。同樣地，一對一式的管理當然有用，但新手主管需要營造一種團隊氣氛，建立一種合作、共襄盛舉式的環境，讓個別部屬投入工作。最後，新領導人需要為團隊的成功創造條件；也就是說，他們需要為團隊撐腰，運用權力與影響

力為團隊爭取利益。

希爾與坎特‧林內貝克（Kent Lineback）以〈你是好老闆──還是偉大的老闆？〉（Are You a Good Boss, or a Great One?）為題，合作寫了一篇論文，談到一些同樣的主題。文中指出，許多老闆由於不求精進，以至於無法充分發揮潛能。這些老闆沒有不斷自問：「我做得多好？」與「我需要做得更好嗎？」這類問題，兩位作者認為，真正知道自己需要做什麼才能確實有效，並了解自己今後志在何方的老闆不夠多。他們提出一套做法，幫助領導人發揮潛能，稱為三大關鍵能力：自我管理、人脈管理、團隊管理。

領導人如果想要成功，就需要影響其他人。追隨者也會觀察自己的老闆，決定是否心甘情願地讓領導人來影響他們，他們必須先信任老闆，才會讓老闆影響他們。如果信任的產生來自能力與個性，領導人就必須管理自己，以展現自己的能力與個性，來取得部屬的信任。

有效的領導人也善於管理自己的人際網絡，他們不會迴避組織政治，反而會盡情擁抱它。因為他們知道，如果想以一種有生產力的方式來發揮影響力，就必須在組織裡

建立適當關係。想取得完成任務所需的資源與權力，最好的辦法就是在組織的各個角落，建立非正式的人脈並投入組織政治。有效的領導人不僅營造、維護這些關係，還會在幾個層面上進行這類工作，例如他們會把自己的老闆也納入網絡中。

在領導一支團隊時，領導人往往禁不住誘惑，比較常與團隊個別成員，而不是團隊整體打交道。時間很珍貴，每個人都在賣力工作，線上團隊會議的效果，有時也不怎麼好。不過，人總是喜歡身為團隊的一員，也喜歡與他人分享共同目標，感受到那種集體使命感。儘管這或許不是最簡單的選項，但有效領導人在管理一支團隊時，會將團隊視為一支隊伍，而不是一群個人。每個人都需要被接納，領導人有時需要針對個別成員進行個別處理，但這類互動總是可以在團隊環境中進行的。

最後，有效領導人必須時時注意自己在三大關鍵能力上的表現。希爾與林內貝克，做了一份清單問卷，協助領導人為自己評分。

▸ 未來你想往哪裡走？

史蒂夫・雷德克里夫（Steve Radcliffe）是領導與發展顧問，曾與聯合利華（Unilever）、英國文官署（Civil Service）等大型組織的執行長並肩工作，著有《領導力：愈簡單明瞭愈好》（*Leadership: Plain and Simple*）一書。

雷德克里夫一直宣揚一個理念：領導未必是一件難事。他認為，無論是跨國企業執行長，或是剛起步的新手主管，大多數人都可以透過學習，成為更有效的領導人。

你在組織擔任什麼職位，一點關係也沒有。你可以是初入職場的新手，沒有任何直接部屬。你可以是團隊領導人，負責一整個部門的運作。你可以是一個組織的負責人，也可以在一所學校、一個慈善團體或一間全球性企業工作。所有這些差異，其實一點關係也沒有，因為我在這些職位上都見過令人鼓舞的領導。我也發現，無論在任何情況，領導的基本原則都是一樣的。

It absolutely doesn't matter where you are in an organization. You can be in your first job, you can have no direct reports. You can have a team or run a department. You can head an organization.

And you can work in a school, charity or global business. It really doesn't matter because I've seen inspiring leadership from people in all these positions and I've realized that the fundamentals of leadership are the same for any situation.

———

雷德克里夫說，無論處在什麼職位，領導人必須注意 3 個不可或缺的要件：未來─投入─交出成果（future-engage-deliver）。領導必須以未來為起始點，領導人必須說明自己未來想往哪裡走。只有強烈知道自己未來想走的方向，領導人才能說服其他人也投入。領導人對未來愈是充滿熱情，帶給追隨者的正面衝擊也就愈大。

接下來，雷德克里夫說，領導人必須懂得投入那個願景。所謂「投入」並不是「與人溝通」、「向人說明」或「告訴人」，而是讓其他人也能懷抱領導人的願景，與領導人一起工作。雷德克里夫表示，想要達到這個目標，領導人必須具備「正直、開放與前後一致」等特質。最後，領導人必須能夠交出成果，或者更精確地說，必須能夠協助團隊，讓團隊幫領導人交出成果。想要改善這 3 項要件，就必須營造、鍛鍊領導資質。我們都有這些資質，問題只在於能不能有效運用而已。

▶ 最懂領導人的教練

如果說這世上只有一個人能夠了解領導人，那這個人大概非葛史密斯莫屬。葛史密斯是全世界最知名的高階主管教練之一，《華爾街日報》（ *The Wall Street Journal* ）將他列為最頂尖 10 大主管教育家。他住在加州，為了替 70 幾位大型組織的執行長擔任教練，搭機風塵僕僕了 7 百多萬哩。他的著作包括《培養領導力》（ *Coaching for Leadership* ）、《未來領導人》（ *The Leader of the Future* ）與《UP 學：所有經理人相見恨晚的一本書》（ *What Got You Here Won't Get You There* ）等。

身為一個領導人，應該在表現上下多少工夫？

這是個非常好的問題。我在與主管們共事時，經常用一齣紐約百老匯或倫敦西區劇院（West End）的故事來做例子。演出這些戲的人不會說：「我的腳好痛；我今天不大舒服；我心情很壞。」為什麼？因為戲在上演了。

我告訴主管們：「舞台上這些孩子賺的錢，只有你賺的 2％。如果連這些孩子都能一天一天又一天，每晚登台像專業人一樣演出，你也可以。」

To what extent is being a leader about putting on a performance?

That is a very good question. The example I use with the executives that I work with is a Broadway or a West End play. People in a show do not say, "Oh, my foot hurts; I don't feel too good today; I'm in a bad mood." Why? Because it is showtime.

I tell the executives, "The kid on the stage is making 2 percent of what you're making. If the kid can go out there, night after night after night, and be a professional, then so can you."

———

但這與「真誠」能接得上軌嗎？

你談到「真誠」。但我不是要主管虛偽，我要他們表現專業。假設你是一家市值百億美元大企業的執行長，你正在開會，會議室裡每個人都望著你的臉。他們仔細聆聽你說的每一個字，你說的每一個字對他們都很重要。

不過，執行長也是人。有時你在開會，有人在那裡做簡報，你感到無聊透頂，因為你早就知道那人要說些什麼，而且你想上洗手間。

但就算再無聊，你還是得正襟危坐、專心聆聽。因為大家都在看你，你若露出不感興趣、不關心或不熱衷的表

情，會讓大家都無精打采。身為專業人士，就不能這麼做。

But how does that tie in with authenticity?

You use the word authentic. But this isn't being a phony; this is being a professional. If you're the CEO of a multibillion-dollar corporation, and you're in a meeting, everyone in that room is looking at your face. They're listening to every word you say, and it matters to them.

Now, CEOs are human. Sometimes you're in a meeting, someone's making a presentation, it's boring, you already know what the person's going to say, and you've got to go to the bathroom.

It doesn't matter. They're all looking at your face, and if you don't look interested and caring and motivated, you demoralize people. That's what being a professional is.

———

你寫的那本《下一步，我該怎麼走？》（*Mojo*）呢？

《下一步，我該怎麼走？》與之前那本《UP 學》非常不一樣。《UP 學》談的是人際關係，《下一步，我該怎麼

走？》談的主要是個人內在的問題。在《下一步，我該怎麼走？》這本書裡，我的討論重心是讓我們快樂、讓我們的生活有意義，我們做什麼才會快樂、才能活得有意義。

我一直在問一個問題：成功的人必備哪些要件？成功的要件很多，但我們可以說，成功的人總是從事能夠「一箭雙鵰」的事。其一，是讓他們快樂；其二，是對他們有意義。

What about your book *Mojo*?

Well it is very different from the previous book, *What Got You Here Won't Get You There*. That book is about interpersonal relationships. *Mojo* is much more intrapersonal. It's on the inside, and in the book *Mojo*, I focus on achieving happiness and meaning in our lives, doing what makes us happy, and doing what is meaningful for us.

I always ask the question, what are the number one characteristics of successful people? One of the key answers is that successful people are engaged in behavior that does two things at once. One, it makes them happy. And two, it's meaningful for them.

——

管理接班程序，是領導人的一項重要角色。你寫過一本
這方面的書──領導人怎麼做才能順利完成接班？

我在書中談到，準備交班的人，需要注意 3 個變數。首
先，當然了，不管你經營的是公司或是事業，你還是得
繼續經營下去。

其次，你當然需要培養接班人。

第三，你需要找其他的事情做，有關這一點的討論很少。

Managing the succession process is an important role for leaders. You wrote a book about that—how do leaders do succession well?

I talk about three variables that the person who is getting ready to leave needs to work on. One is, of course, that you still have to run your company or your business, whatever it happens to be.

Two, you need to develop your successor, of course.

Three, which is seldom discussed, you need to find something else to do.

──────

所以說，這本書以非常實際的角度觀察領導接班？

這本有關接班問題的書，之所以讓我得意，是因為它討論接班問題的現實面。它不裝模做樣、標榜什麼長期股東價值議題，而以人性面為討論重點。接班給人什麼感覺？把一切事情交出去，放下你做了一輩子的工作是什麼滋味？

當然，現實是，那非常難，非常、非常難。

以接力賽跑為例。接力賽跑很不容易。如果你領先，大家都會替你喝采、加油，要你繼續跑、不要停。如果你落後，也不肯停下來。你會不斷告訴自己，我非趕上不可，我非趕上不可。

所以，無論你跑在前面或後面，要你放手交棒都很難。在《交班：你準備好了嗎？》（*Succession: Are You Ready?*）這本書裡，我討論了放手的動能，以及它何以重要。我也談到交班之難。我開過 3 次課，討論放手交班的問題，每堂課有 11 位執行長參加。我可以告訴你，接班這件事說起來容易，做起來難。

So it is a very practical look at this aspect of leadership?

What I like about this succession book is that it talks about the reality of succession. Rather than pretending that this is a process in which everybody talks about long-term shareholder value, it covers the human dimension. What does succession feel like? What does it feel like to let go? What does it feel like to give up what you're doing?

Of course, the reality is that it's hard. It is very, very hard.

Take a relay race, for example. It's tough. If you're ahead in the relay race, everybody's cheering, keep going, don't stop. And if you're behind, then you don't want to stop. You feel like, I've got to catch up, I've got to catch up.

So either way, it is hard to let go. In the book *Succession: Are You Ready?*, I talk about the dynamics of letting go, and why it's important. I also talk about how hard it is. I've done three sessions, with 11 CEOs in each session, talking about letting go. I can tell you, it's easy in theory, but not easy in practice.

———

彼得・杜拉克曾說，我們花太多時間告訴領導人他們應該做什麼，卻花太少時間在告訴他們應該不做什麼。你同意這個觀點嗎？

完全同意。彼得・杜拉克說，我們花太多時間幫領導人學習應該做什麼，卻沒有花足夠時間幫他們學習不做什麼。

事實上，杜拉克這句話，就是我寫《UP 學》的靈感。這本書許多內容談的，就是教成功領導人怎麼放手。

有許多次，當我們在為領導人授課的時候，提出的建議並不深奧，講的話也很簡單：不要再做下去了。如果大家能夠學會「不做」什麼事，就會變得好一點。比方說，碰上一些頑固、有成見的領導人，我會告訴他們，在與我說話的時候，每句話不得用「不」、「但是」或「可是」這幾個字開頭。每用一次這些字，我就罰他們20 美元。

某天，在與一位客戶進行 360 度反饋評量時，他說：「但是，馬歇爾」，我立即對他說：「這次放你一馬，但如果你和我說話時，再用『不』、『但是』、『可是』這幾個字開頭，我就要罰你 20 美元。」他立刻回答：「但是，馬歇爾」，結果便交出 20 美元。然後他又說了「不」：40 美元；之後再說 3 個「不」：60、80、1 百美元。我們談了一個半小時，他被我罰了 420 美元。在這一個半小時的課程結束時，他對我說：「謝謝你。我想不到會像這樣。有你當面提醒，我還犯了 21 次。如果不是你當面提醒，我會犯多少次？50 次？1 百次？難怪許多人認為我頑固、堅持己見了。」

It was Peter Drucker who said that we spend too much time telling leaders what they should do, and not enough time telling them what they should not do. Do you agree with that?

Totally. Peter Drucker said that we spend a lot of time helping leaders learn what to do, and not enough time helping leaders learn what to stop.

That quote was actually the inspiration for my book *What Got You Here Won't Get You There*. A lot of that book is about teaching successful leaders what to stop.

A lot of times, when we coach people, what we tell them isn't deep or profound, it's simple: quit doing this. And if people learn to quit doing things, they get better. For example, if you're stubborn and opinionated, I teach people, don't start sentences with no, but, or however. I fine my clients $20 every time they do it.

So I'm going over one client's 360-degree feedback, and he says, "But Marshall," and I say, "It's free this time, but if I talk to you again and you start a sentence with no, but, or however, I'm going to fine you $20." He says, "But Marshall": $20. "No": $40. "No, no, no": $60, $80, $100. He lost $420 in an hour and a half. At the end of an hour and a half, he said to me, "Thank you. I had no idea. I did that 21 times with you throwing it in my face. How many times would I have done it if you had not been thrown it in my face? 50? 100? No wonder people think I'm stubborn and opinionated."

_____

你提到「謝謝你」這幾個字，這些字在領導過程中重要嗎？

非常重要。而且它們的重要性，並不會因為職位變高而減少。職位愈高，它們對你更重要。

每個決定都是由有權決策者所做的，不是由那個最聰明、最好或最對的人所做的。你的職位愈高，便愈有可能成為那個有權做決定的人。你必須贏，你必須一直贏。要領導人放棄這種證明他們聰明、他們對或他們好的意願，並感謝其他人、將貢獻歸於其他人，非常難。

但是，你應該讓其他人聰明，應該讓其他人對、讓其他人贏。如果你是執行長，無論如何你都會贏，不需要把功勞全部攬在自己身上；你需要把功勞讓給其他人。

我的看法是，你愈是能夠把好點子歸功於其他人，愈是能夠為其他人的貢獻向他們致謝，愈是能夠讓其他人因為對組織的貢獻而得意，便愈能夠皆大歡喜。

You mentioned the words thank you. How important are those words in leadership?

Very important. And they don't become less important the higher up you go. They become more important.

Every decision is made by the person who has the power to make that decision, not the smartest person or the best person or the right person. The higher up you go, the more you are that person. You get to win. You get to win all the time. It is very hard for leaders to let go of this desire to prove that they're smart, or right, or good, and to thank other people, to recognize the contributions of others.

But let the other person be smart. Let the other person be right. Let the other person win. If you're the CEO, you get to win anyway. You don't need to take credit; you need to give credit.

To me, the more you can let other people take ownership of ideas, thank them for their contributions, and make them feel good about what they're doing for the organization, the better off everyone is.

———

我們談的一直是領導，告訴我們一點關於你的教練方法吧。

我的職責是協助已經成功的領導人，讓他們能夠持之久遠地改善自己的行為、改善他們團隊中其他人的行為。所以，我做的是非常特定、非常專注的事。我的任務不在於解決問題，而在於協助已經非常成功的人，他們都已經很了不起，而且想做得更好。

我的教練方法很獨特。如果我的客戶不能變得更好，我什麼報酬也拿不到。而且他們是不是變得更好，並非我說了算，也非我的客戶說了算，評斷這位客戶是否變得更好的，是這位客戶身旁的每個人。

We have been talking about leadership, but tell us a little bit about your coaching methods.

My job is to help leaders who are already successful achieve positive, lasting change in behavior for themselves and the people on their teams. So what I do is very specific and focused. It's about not fixing problems, but about helping people who are already very successful, great people try to get better.

My coaching method is unique. I don't get paid if my clients don't get better, and whether they get better is not judged by me or by my clients. It's judged by everyone around my clients.

——

這麼說，你是否曾經白忙一場，結果拿不到錢？這種事發生過嗎？

發生過。但並不多；大約有 10% 或 15% 的教練時間，基於各種理由我沒有得到酬勞。不過，我的教練運作方式非常直截了當。

接受我教練的人，必須得到其他每個人如何看待他的祕密反饋。他要找出自己哪些事做得很好、哪些地方需要改進。他身旁的人要提出建議，之後他與我坐下來，有時他的老闆也會與我們一起坐下來討論。我們必須達成一個協議。他必須取得其他人對他的反饋，與其他人討論，並且定期、有紀律地進行後續檢討，為過去犯的錯致歉。

我的合約很簡單：你要比過去好。如果由正確的人做出評斷，認為你的行為對了，那這些錢花得很值得。我也會告訴他們，如果這些錢花得不值得，那就不要做。如果這些錢花得值得，你怎麼樣都是贏家。你比過去好，我拿到報酬。你不比過去好，我的教練免費。

So how often do you not get paid? Has it ever happened?

It's happened. Not a lot; 10 or 15 percent of the time I don't get paid for a variety of different reasons. But the way my coaching process works is very straightforward.

The person who is receiving the coaching has to get confidential feedback on how everyone sees him. He is going to find out what he's doing well and what he needs to improve. The people around him venture suggestions. Then he and I sit down, possibly with his boss, and talk. We have to reach an agreement. He's going to have to get the feedback, talk to people, follow up on a regular, disciplined basis, and apologize for previous sins.

And my contract is simple: you get better. If the right behavior is judged by the right people, it's worth this money. And I tell the people, if it's not worth the money, don't do it. If it is worth the money, you can't lose. You get better, I get paid. You don't get better, it's free.

———

那個「前饋」（feedforward）要件談的是什麼？能與我們談談嗎。

是這樣的，我是個佛教徒，一個哲理上的佛教徒。「前饋」是一種非常佛教的概念。在教「前饋」的過程中，我對大家說，不要徵求關於過去的反饋；要徵求關於未來的想法。

我教我的客戶，要他們閉上嘴、傾聽、做筆記，並且要說謝謝。無論對方說的是什麼，你都要向對方說謝謝。不要向對方保證，你會做到他們提出的一切要求。領導不是在比賽人氣。我的客戶只要聆聽、細想別人說些什麼，記下來，然後盡可能行事。

「前饋」的重點在於未來，你可以改變未來，不能改變過去。這與貶低任何人或侮辱任何人，一點關係也沒有。「前饋」的過程中不容許裁判，所以它很正向、很樂觀。大家都很喜歡這種構想，因為「前饋」可以取得約 80％ 反饋之利，卻不需要付出憤怒、貶低他人、為自己辯解的一切成本。我愛「前饋」這種構想；它是我培訓過程的基本要素。

And what about the *feedforward* element? Tell us a little bit about that.

Well I'm a Buddhist, a philosophical Buddhist. *feedforward* is a very Buddhist concept. In *feedforward*, I teach people, don't ask for feedback about the past; ask for ideas about the future.

And I teach my clients to shut up, listen, take notes, and say thank you. No matter what the person says, just say thank you. Don't promise to do everything that people suggest. Leadership is not a popularity contest. My clients just listen and think about what people say, write it down, and then they do what they can.

It's focused on a future that you can change, not a past that you can't change. It doesn't involve putting anyone down or insulting anyone. No judging is allowed, so it's very positive and upbeat. People like it, and you get about 80 percent of the benefit of feedback. You miss all the cost of the anger, the putting people down, and the defensiveness. I love the *feedforward* idea; it's the essence of my coaching process.

———

如何才能做到像你這樣優秀的主管教練？需要什麼資質？你從你的工作中，學到了什麼？

你知道嗎，依我看，最重要的資質就是放下你的自我。如果我必須回顧自己身為教練這一生遭遇的種種失敗，頭號的失敗就是我自己。

經過我教練之後而改善的客戶，一般說來，是我在時間上花得最少的客戶。經過我教練之後絲毫未見改善、也讓我白忙一場的客戶，一般說來，是我在時間上花得最多的客戶。

某次，我對一位客戶說：「我應該從你身上學到什麼？」他說：「馬歇爾，大概可以學到兩件事吧。身為教練，你需要學的第一件事就是慎選客戶。選對了客戶，你的教練做法百戰百勝。若找錯了客戶，你怎麼教，都是白費苦心。我的工作也沒多大不同。我要管理許多很了不起的人。身為領導人，我有多能幹其實不要緊。我如果用人不當，怎麼做也無法成功。」

然後他說：「第二件事就是，不要把教練過程的重心放在你身上。這事的重點不在你，也不在你的自我；它的重點在我、在我的團隊。不要迷失於自我。」接著，他補充一句：「我的工作也一樣。如果是談高成就者，那重點在我。但如果是談卓越的領導者，重點則在他們身上。」

身為教練，我學到一個很不容易學到的教訓，那就是想當一位了不起的企業教練，就必須拋開自己，因為一切重點都在客戶。我學到身為教練必須知道的一件事，那就是教練工作的一切不在於我，這是一個很不容易學到的教訓。

And what makes a good executive coach like yourself? What are the qualities? And what have you learned from what you do?

You know, I think the biggest quality is letting go of your own ego. If I had to look at my failures in life as a coach, the number one failure would be me.

The clients I coached that improved are the clients that I spent the least amount of time with. The clients I coached that didn't improve at all, so I didn't get paid, are those that I spent the most amount of time with.

I said to one of my clients, "What should I learn from you?" He said, "Marshall, a couple of things. The first thing you need to learn as a coach is that your number one job is client selection. If you have the right clients, your coaching process will always work. If you have the wrong clients, your coaching process will never work." And he said, "My job isn't that different. I have to manage great people. It doesn't matter how good I am as a leader. If I have the wrong people, I'm not going to be successful."

He then said, "The second learning point is, don't make the coaching process about you. This wasn't about you and your ego; it was about me and my team. Don't get lost in yourself." He added, "My job is the same. As a great achiever, it's all about me. As a great leader, it's all about them."

The hard lesson for a coach as a great coach is that it's not about you, it's about your clients. And the one thing I've learned as a coach, a very hard lesson, is that it's not about me.

———

你有一套「同儕教練」（peer coach）程序，裡面有一套問卷。你能告訴我們一些問卷內容嗎？

我可以說幾個特定問題。不過，這套程序是這樣的：我做問卷，做給我自己回答。你得自己寫問題問自己，這是「同儕教練」的構想。

我說幾個我為自己寫下的問題，或許有人會有興趣。我每天都要回答一個問題：「從 1 到 10，你昨天有多快樂？」我不需要工作，我住的地方很好，也有好朋友、愛護我的家人，還有很好的客戶。如果我還不快樂，這是誰的問題？照照鏡子吧。

「從 1 到 10，你昨天過得多有意義？」我是否做了具有關鍵性、重要的事？還是一整天都浪費了？

「你曾有多少次設法證明自己是對的，但其實根本就不值得這麼做？」我真不願意承認，但我這輩子在回答這個問題的時候，幾乎從來沒有答過一次「零」。想不證明自己是對的，真的很難。

「你對其他人說了多少氣話，或具有破壞性的話？」
「你有沒有對妻子、兒女說或做一些好事？」

「你花了多少時間寫作？」我不知道。我已經寫了 28、
29 本書。這些書不會自己寫出來，你得一個字、一個字
寫才行。

「你不斷更新客戶資料嗎？」「你的體重有多重？」
「你喝了多少酒？」這些都只是人生基本問題，我發現
問自己這些問題，能夠讓我保持專注。

有人曾經問我：「你做這個幹嘛？你沒聽過改變行為的
理論嗎？」

那個理論是我寫的。我做這些事，為的就是這個。

**You have a process with your peer coach where you go through a
series of questions. Can you reveal what some of those questions
are?**

Well, let me describe some of the specific questions. And the way
the process works, by the way, is that my questions are intended
for me. The idea of the question process is that you write your
own questions.

I'll share some of mine, though, just in case others might be interested. The first question every day is: "On a 1 to 10 scale, how happy were you yesterday?" I don't have to work. I live in a nice place. I have nice friends and family and wonderful clients. If I'm not happy, whose problem is that? Look in a mirror.

"On a 1 to 10 scale, how meaningful was yesterday?" Did I do something that mattered, something that was important, or did I just waste time?

"How many times did you try to prove you were right when it wasn't worth it?" I hate to say this, but I've almost never in my life gotten zero. It is hard not to do this.

"How many angry or destructive comments did you make about other people?" "Did you say or do something nice for your wife, your son, or your daughter?"

"How many minutes did you write?" I don't know. I've written 28 or 29 books. They don't write themselves. You actually have to do the work.

"Are you updated on your clients?" "How much do you weigh?" "How many alcoholic drinks did you have?" They're just basic questions about life, and I find that this keeps me focused.

Someone once asked me, "Why do you need to do this? Don't you know the theory about how to change behavior?"

I wrote the theory. That's why I do it.

————

▶ 如何知道自己跟對老闆？

　　像是為領導人做一個總結一樣，《領導力密碼：領導的 5 條守則》（*The Leadership Code: Five Rules to Lead By*）這本書很有說服力。本書共同作者凱特・史威曼（Kate Sweetman），是麻省理工學院列格坦發展與創業中心（Legatum Center for Development and Entrepreneurship）訪問學者，她在這個中心與來自新興和開發中國家極有才華的青年創業家共事。

追隨者可以做什麼，確定自己跟到的是好領導人？

事實是，組織需要很多領導人，在組織裡追隨者總是比領導人多。

如果你是從領導力，來考慮自己喜不喜歡在一個地方工作，領導力密碼這5條守則，是一個還不錯的切入點。

What can followers do to make sure that they get good leaders?

The fact is, organizations need a lot of leaders, but they are probably going to have more followers than leaders.

The five elements of the leadership code are actually not a bad list to tick through if you're thinking about whether a place is somewhere you might like to work, from a leadership perspective.

———

應該自問一些什麼樣的問題？

舉例來說，首先你可以自問，這個地方的決策層主管，是不是真的知道他們要做些什麼。他們是否擁有將組織帶向未來的遠見、使命感、策略等。

其次，這些主管看起來是否知道如何執行那些策略？這個地方的工作，看起來是不是進展順利？產品順利上市嗎？該發生的事，是否都發生了？

第三，他們真的知道如何與自己的員工產生連結？組織往往只顧執行，忘了推動這一切工作的，其實都是有血有肉的人，特別在處境艱難時更是如此。這個地方的情緒脈動如何？給人什麼感覺？在這裡工作的人，看起來是不是很興奮、很高興來這裡上班？他們是不是認為這

裡的工作有前途？還是說，這裡的人無精打采，因為他
們實在不覺得在這裡工作有什麼發展。

What kinds of questions should you ask yourself?

For example, begin by asking whether you think that the people
who are at the decision-making level really have a good sense
of where they are going. Do they have a vision, a mission, a
strategy, or whatever they are calling it, about how they are going
to translate the organization into the future?

Second, do they appear to know how to execute on that strategy?
Do things seem to work smoothly; do their products get to
market; do things really happen?

Third, do they really know how to connect to their own people?
Often, particularly in these difficult times, organizations are
so focused on execution that they forget that there are human
beings behind the organization. What is the feel of the place, the
pulse? Do people seem to be excited to come to work and happy
to be there, and do they feel like there is a future for them, or is
the tone around the place more that it is not really a place where
you are going to be able to move forward?

———

還有其他要注意的嗎？

進組織見到老闆以後，你對這個人有什麼感覺？這個人是不是你可以信任、依靠的對象？這個人能夠照顧你，同時也照顧自己嗎？

所以說，在投入工作以前，你需要做一番複雜的評估，但我認為這麼做是值得的。

Is there anything else?

Well, when you get in to meet the individual you are going to work for, what is your sense of him or her as a person? Is this really somebody whom you can rely on and trust, who you think is going to be looking out for you as well as for his or her own interests?

So, there is a sort of complex assessment that you need to make, but I think it is a worthwhile one.

———

你們寫這本書，一開始的構想是什麼？

我們之所以寫《領導力密碼》，為的不是創一套新領導模式，也不是提出革命性的領導思考方式；我們只是想為非常令人困惑的領導世界，帶來一些秩序。

我在領導這個議題上工作了 20 年。我觀察過許多組織，發現組織與組織部門不同，對領導要件的強調也不同。例如，有些組織與部門強調 EQ，有些特別重視調適領導或情境領導等。

但這些組織、組織部門，有沒有什麼共同準則？我們寫這本書的目的，就是要找出領導力的共同準則，找出各式各樣領導理念的切入點。幫助看過這本書的讀者，能以一種周詳而平衡的方式選擇當一個領導人，或選擇在一個組織裡培養領導能力。

What was the idea that you set out with?

The reason we wrote *The Leadership Code* was not so much to come up with a new model of leadership or a revolutionary new way to think about it; it was really more about trying to make some order of the very confusing universe of leadership.

I have been working in leadership for 20 years. I have been in many organizations, and it is obvious that organizations, and different parts of organizations, choose different elements of leadership to emphasize. It might be emotional intelligence, or adaptive leadership, or situational leadership, for example.

But what does it all really add up to for these organizations?

What we have tried to do is to look at what all leadership has in common and how these different ideas really fit in, so that when you are choosing to become a leader or to develop leadership in an organization, you know that you have thought about it in a thorough and balanced way.

———

好的領導，有共同要件嗎？

絕對有。比方說，乍看之下，德雷莎修女（Mother Teresa）與邱吉爾幾乎沒有什麼共同點。但事實上，你若進一步思考，就會發現所有偉大的領導人，都能夠建立歷久不衰的組織。觀察他們的所做所為，你會發現他們做的事，有 60％ 到 70％ 是一樣的。

Are there common elements to good leadership?

Absolutely. On the face of it, you would think that Mother Teresa had very little in common with Winston Churchill, for example. But, in fact, when you consider them more closely and you look at good leaders who could really build organizations that lasted, in all sorts of endeavors, what you find is that 60 to 70 percent of what they do is actually the same.

———

如果這 5 條守則，涵蓋有效領導要件的三分之二，其他還有什麼要件？

其他要件主要視情況而定，例如，你是想用某種方式經營一家製藥公司，還是想整列大軍、準備打仗等。有些領導人在一個環境下做得有聲有色，換一個環境就不行了，原因就在這裡。

If the five rules account for two-thirds of what makes a leader effective, what else is important?

The other components really depend on the situation—on whether you are trying to run a pharmaceutical company in a certain way or whether you are trying to marshal an army, for example. That's why a leader who is successful in one endeavor sometimes is not so successful in another.

———

妳能詳細說明領導力密碼的 5 條守則嗎？

領導力密碼的 5 條守則，其實涉及長程與短程，也涉及業務與人。

以一個典型的四方塊顧問矩陣為例，右上角是擅長長程思考，而且非常關注業務的人，所以是「策略人」

（strategist）。「策略人」這塊的領導力，在於了解較廣的景象、掃瞄環境，並與組織外部的參與者（特別是顧客）談話。它的重點在於從組織內部觀察外在世界，以便了解你能與組織一起走向何方。

領導人有時會忘記我們稱為「策略磨擦」（strategic traction）的一項策略要件。在訂定策略時，你要記住一件事：策略需要一個執行它的組織。所以說，假設一項策略的一部分是擴張，你得將組織具備多少能力、可以擴張到什麼程度的問題，也納入策略考量。

Can you elaborate on the individual elements of the leadership code?

The five rules of the leadership code really have to do with the long and the short term, and with the business and the people.

So, if you think about a classic four-square consulting matrix, in the upper right-hand corner is someone who thinks about the long term and is really focused on the business, so that is the strategist. Thus, the strategist piece of leadership is really about understanding the larger picture, scanning the environment, and talking to external stakeholders—in particular, customers. It is also about really understanding, from the inside of the organization, what you know about the world and therefore where you can go with the organization.

Another important piece of strategy that leaders sometimes forget about is what we call strategic traction. As you formulate the strategy, remember that there's an organization that needs to execute on it. So while part of the strategy formulation is possibly about stretch, it also has to include the organization's capability to deliver against that.

———

接下來是什麼？

我們稱領導力密碼的第二部分為「執行人」（executor）。執行人專注於業務，能夠進行短程行動。有人將這一塊稱為管理，但管理其實是領導的一部分，是領導之所以能成為領導的主因。

組織準備改變計劃，訂了一套改變辦法，領導人知道這些決定怎麼下達，也知道由誰來推動這些決定。這時，你就必須應付各種團隊議題，務使所有相關人等都能群策群力，根據策略來推動計劃。

What comes next?

The second part of the leadership code we call the executor. The executor is focused on the business, but is able to act in the short term. It is what some people might call management, but it is the part of leadership that is really about making it happen.

So there is a change plan in place and a methodology for going about it, and the leader knows how the decisions are going to be made and who's going to do it. This is where you work through a lot of those team issues to make sure that everyone is put together in a way that will move the plan forward, tied to the strategy.

———

第三部分呢？

矩陣左下角，是人與短程目標相會的地方。所以，如果我們是領導人，就必須能夠與身旁的這些人「溝通」，讓他們知道究竟該怎麼做。

舉例來說，在充滿不確定性的艱難時刻，只怕溝通不夠，幾乎不可能太過。我們常在組織裡看到這一幕：老闆訂了一個奇特的策略，或想出一套帶動組織的妙方，興高采烈揚長而去，但員工卻一片茫然，完全不曉得發生了什麼事。他們真的很需要有人與他們溝通。

這一塊的要旨，也涉及與你的部屬聯繫，了解他們的動機，管理他們的才幹。

How about the third part?

The box in the bottom left is where the people element meets the short term. So, if we are leaders, we have got to be able to communicate with the people who are surrounding us right now, letting them know exactly what it is required.

At times when people are living through a lot of difficulties and uncertainties, for example, it is almost impossible to overcommunicate. Often we see in organizations that when the boss is off planning some exotic strategy or coming up with a method for the organization to go forward, the employees really need to have someone talk with them because they need to know what's going on.

This dimension is also about connecting with your people, understanding their motivation, and managing the available talent.

———

那第四個層面呢？

這是大多數組織最忽略的一塊：「人力資本發展」

（human capital development）。這是商業策略與人長程接軌的一塊。所謂「人力資本發展」的重點，就在於領導人在決定將組織帶到什麼地方去以後，就要考慮應該如何組織團隊？就區域範圍而論，應該選在哪裡落腳、營運？哪些職位具有關鍵性？知不知道誰適合擔任這些職位？等等。

And the fourth dimension?

This is the part that most organizations miss the most: human capital development. It is where strategy concerning the business meets people in the long term. So, what human capital development is really about is, given where the leader thinks the organization is heading, how is it going to be organized? Regionally, where will it be located and operating? What are the key jobs going to be? Do we have a sense of who is going to be able to fill those kinds of jobs? And so on.

——

這是領導人的 4 個外向層面，領導人還必須具備第五項要件，就是對自己身為領導人的自我了解，對嗎？

是的。這第五個面向談的，就是領導人的本源。我們可以將它視為一種「個人能力」（personal proficiency）的層

面。學界對這個領域已有諸多探討，報刊與媒體對這個
領域也已有很多報導。例如 EQ 和真誠領導，就是屬於這
個領域的要件。

但我們所謂的「個人能力」，基本上指的是從你身上應
運而生、能幫你做好其他事情的能力，如它能幫你做
個「策略人」，能幫你做個「執行人」，能幫你與他人
打交道，能幫你規劃組織未來發展，還能幫你了解組織
今後會需要哪些關鍵人才等。所以說，所謂「個人能
力」，重點在於領導人必須知道他們想達成什麼。

**Those are the four external dimensions of a leader, then, plus
there is also a fifth element of understanding yourself as a leader.
Is that right?**

Yes. This dimension is about the source of the leader. Think of
it as the personal proficiency dimension, really. It is an area that
there has been a lot of work on, and where there has been a lot
of coverage in the press and in the media. So, for example, it is
the dimension where you would be including elements such as
emotional intelligence and authentic leadership.

But what we call personal proficiency is basically the things
that emerge from you as a person that are able to help you to
do these other jobs well: being a strategist, being an executor,

dealing with people, and figuring out the map of the future and the people map of the future. So, when you think about personal proficiency, it really has to be about leaders understanding what it is that they want to accomplish.

——

▶ 領導很單純，就是有關人的事

總結而論，領導其實很單純。但在一個複雜的組織裡，日復一日地領導，是非常艱鉅的工作。貝恩管理顧問公司（Bain & Company）的合夥人克里斯‧祖克（Chris Zook），談到一種他所謂「指揮官的意圖」（commander's intent）的概念。下列就是他針對這個概念，對我們提出的解釋：

「指揮官的意圖」這個概念，早自 1700 年代與 1800 年代已經萌生，經多次海戰與拿破崙戰爭淬鍊而逐漸成形。這項概念是這樣產生的：拿破崙戰爭期間英國海軍提督納爾森上將（Admiral Nelson）等人發現，在面對一種情勢時，如果能使用一種非常簡單、讓每個指揮官都知道、都了解的聲明，這項聲明能發揮很大的威力。

這是一種策略聲明,聲明中還會陳述一些不容談討價還價、沒有妥協餘地的行為原則,因為戰鬥過程中總會出現一些難以預期的事件,推到第一線的決策愈多愈好。

納爾森上將在海上幾乎戰無不勝,是公認非常成功的領導人。他之所以能夠這麼成功,部分原因是他與一群親密戰友,也就是麾下其他艦長,有一套相當明確的行為原則。這些艦長就算遇上始料未及的狀況,例如就算遠在海平線另一端、看不到納爾森,也能預知其他戰友會採取什麼行動。

在商務上,所謂「指揮官的意圖」,就是不容妥協的商業原則。許多最成功、基業最持久的公司,都是能將決策點盡量逼近第一線、能將組織中間層級盡量簡化的公司。這是因為公司上下都能清楚了解「指揮官的意圖」——也就是公司想做什麼——也都能清楚了解公司不容妥協的商業原則所致。

以全球最大投資公司先鋒(Vanguard)為例。大型投資人是投資界最有利可圖的區塊,也是大多數投資公司必爭的市場。但先鋒極度重視小型投資人,對投資大戶反倒不甚重視。它以互惠結構與指數型基金操作方式強調低成本,因為它認定若是沒有內部資訊,小型投資人不可

能持續擊敗市場，而這項理念已經成為它不容妥協的一項原則。在先鋒公司，上自執行長、下自講電話的客服代表，都能將這些原則與聲明說得清楚明白。也因為如此，先鋒能將更多行動、更多學習、更多決策推向第一線，讓眾多省思與見解以更有秩序的方式回饋給公司。總結而論，領導是一種多面向、講究平衡的行為。它涉及原則，它需要行動。它迫使我們學習。它既受環境與時勢所驅，也必須仰仗個人的活力與抱負。它是非常有人味的東西。

Over time, actually tracing back to the 1700s and 1800s even, through naval combat and Napoleonic wars, the phrase commander's intent emerged. It came from a situation where a number of people, such as Admiral Nelson and others, found it very powerful to be able to have a very simple statement that every one of the commanders knew and understood.

It was a statement of the strategy and some of the nonnegotiable principles of behavior, because unexpected elements always emerge during combat, and the more decisions that you can push down to the front line, the better.

Admiral Nelson is believed to have been so successful, winning so many naval engagements, in part because he and his band of brothers, the other captains, had a relatively clear set of principles of behavior. This meant that his captains, even under unexpected conditions, even when they were over the horizon and unable to see Nelson, for example, could almost anticipate how the others would behave.

In business, the analogy is the nonnegotiable principles of the business. Many of the most successful and enduring businesses are those that have actually been able to push decisions closer to the front line, with fewer layers in the middle. This is because the commander's intent, the essence of what the business was trying to do, and the key nonnegotiable principles were so well understood.

Take the example of Vanguard, the biggest investment company in the world. Vanguard has obsessively focused on the small investor, rather than the big investor, which is the more lucrative segment that most companies focus on. It has focused on being low cost through its mutual structure and expenses, and on index funds because it believes, as one of its nonnegotiable principles, that the small investor cannot beat the market consistently without inside information. These types of principles and statements are as clearly stated by the CEO as they are by the person on the phone speaking to the customer. That allows much more action, more learning, and more decisions that can be pushed down to the front line, and also for insights to be fed back in a more orderly way. In the final analysis, leadership is a multifaceted balancing act. It involves principles. It requires action. It demands learning. And it is driven by context as much as personal energy and ambition. It is dauntingly human.

———

國家圖書館出版品預行編目資料

50 大商業思想家論壇 當代最具影響力 13 位大師
談領導 / 史都華·克萊納 (Stuart Crainer), 德·迪
樂夫 (Des Dearlove) 作;譚天譯. -- 初版. -- 臺北
市:麥格羅希爾, 2014.06
面; 公分. -- (經營管理;BM207)

譯自:Thinkers50 Leadership

ISBN 978-986-341-112-3(平裝)

1. 企業領導 2.組織管理

494.2 103007827

經營管理 BM207

50 大商業思想家論壇 當代最具影響力 13 位大師談領導

作　　　　者	史都華‧克萊納（Stuart Crainer）
	德‧迪樂夫（Des Dearlove）
譯　　　　者	譚天
特 約 編 輯	邱慧菁
企 劃 編 輯	廖姿菱
行 銷 業 務	曾時杏　蒙捷中
業 務 經 理	李永傑

出 　版 　者	美商麥格羅希爾國際股份有限公司台灣分公司
地　　　　址	台北市10044中正區博愛路53號7樓
網　　　　址	http：//www.mcgraw-hill.com.tw
讀 者 服 務	Email:tw_edu_service@mheducation.com
	Tel: (02) 2383-6000　Fax: (02) 2388-8822
法 律 顧 問	惇安法律事務所盧偉銘律師、蔡嘉政律師
劃 撥 帳 號	17696619
戶　　　　名	美商麥格羅希爾國際股份有限公司台灣分公司

亞 洲 總 公 司	McGraw-Hill Education (Asia)
	1 International Business Park #01-15A, The Synergy Singapore 609917
	Tel: (65) 6868-8185 Fax: (65) 6861-4875
	Email: mghasia_sg@mcgraw-hill.com

製 　版 　廠	信可印刷有限公司	02-2221-5259
電 腦 排 版	林婕瀅	0925-691-858

出 版 日 期	2014 年 6 月（初版一刷）
定　　　　價	360 元
原 著 書 名	Thinkers50 Leadership

ISBN：978-986-341-112-3

美商麥格羅希爾愛護地球，本書使用環保再生紙印製

10044
台北市中正區博愛路 53 號 7 樓

美商麥格羅希爾國際出版公司
McGraw-Hill Education(Taiwan)

麥格羅・希爾
全球智慧中文化
www.mcgraw-hill.com.tw

感謝您對麥格羅‧希爾的支持
您的寶貴意見是我們成長進步的最佳動力

姓 名：＿＿＿＿＿＿＿＿＿＿ 先生 小姐 出生年月日：＿＿＿＿＿＿

電 話：＿＿＿＿＿＿＿＿＿ E-mail：＿＿＿＿＿＿＿＿＿＿

住 址：＿＿＿＿＿＿＿＿＿＿＿＿＿＿＿＿＿＿＿＿＿＿＿＿＿

購買書名：＿＿＿＿＿＿＿＿ 購買書店：＿＿＿＿＿＿ 購買日期：＿＿＿＿

學 歷： □高中以下（含高中）□專科 □大學 □碩士 □博士

職 業： □管理 □行銷 □財務 □資訊 □工程 □文化 □傳播
□創意 □行政 □教師 □學生 □軍警 □其他 ＿＿＿＿＿＿

職 稱： □一般職員 □專業人員 □中階主管 □高階主管

您對本書的建議：

內容主題 □滿意 □尚佳 □不滿意 因為 ＿＿＿＿＿＿＿＿＿＿

譯／文筆 □滿意 □尚佳 □不滿意 因為 ＿＿＿＿＿＿＿＿＿＿

版面編排 □滿意 □尚佳 □不滿意 因為 ＿＿＿＿＿＿＿＿＿＿

封面設計 □滿意 □尚佳 □不滿意 因為 ＿＿＿＿＿＿＿＿＿＿

其他 ＿＿＿＿＿＿＿＿＿＿＿＿＿＿＿＿＿＿＿＿＿＿＿＿＿

您的閱讀興趣：□經營管理 □六標準差系列 □麥格羅‧希爾 EMBA 系列 □物流管理
□銷售管理 □行銷規劃 □財務管理 □投資理財 □溝通勵志 □趨勢資訊
□商業英語學習 □職場成功指南 □身心保健 □人文美學 □其他 ＿＿＿＿

您從何處得知 □逛書店 □報紙 □雜誌 □廣播 □電視 □網路 □廣告信函
本書的消息？ □親友推薦 □新書電子報 促銷電子報 □其他 ＿＿＿＿＿＿

您通常以何種 □書店 □郵購 □電話訂購 □傳真訂購 □團體訂購 □網路訂購
方式購書？ □目錄訂購 □其他 ＿＿＿＿＿＿＿＿＿＿＿＿＿

您購買過本公司出版的其他書籍嗎？ 書名 ＿＿＿＿＿＿＿＿＿＿＿＿＿＿

您對我們的建議：

＿＿＿＿＿＿＿＿＿＿＿＿＿＿＿＿＿＿＿＿＿＿＿＿＿＿＿＿＿＿＿＿

＿＿＿＿＿＿＿＿＿＿＿＿＿＿＿＿＿＿＿＿＿＿＿＿＿＿＿＿＿＿＿＿

＿＿＿＿＿＿＿＿＿＿＿＿＿＿＿＿＿＿＿＿＿＿＿＿＿＿＿＿＿＿＿＿

＿＿＿＿＿＿＿＿＿＿＿＿＿＿＿＿＿＿＿＿＿＿＿＿＿＿＿＿＿＿＿＿

信用卡訂購單

(請影印使用)

我的信用卡是☐VISA ☐MASTER CARD（請勾選）

持卡人姓名：　　　　　　　　信用卡號碼（包括背面末三碼）：

身分證字號：　　　　　　　　信用卡有效期限：　　　年　　　月止

聯絡電話：（日）　　　　　　（夜）　　　　　手機：

e-mail：

收貨人姓名：　　　　　　　　公司名稱：

送書地址：☐☐☐

統一編號：　　　　　　　　　發票抬頭：

訂購書名：

訂購本數：　　　　　　　　　訂購日期：　　　年　　　月　　　日

訂購金額：新台幣　　　　　　元　持卡人簽名：

書籍訂購辦法

郵局劃撥
戶名：美商麥格羅希爾國際股份有限公司台灣分公司
帳號：17696619
請將郵政劃撥收據與您的聯絡資料傳真至本公司
FAX：(02)2388-8822

信用卡
請填寫信用卡訂購單資料郵寄或傳真至本公司

銀行匯款
戶名：美商麥格羅希爾國際股份有限公司台灣分公司
銀行名稱：美商摩根大通銀行　台北分行
帳號：3516500075
解款行代號：0760018
請將匯款收據與您的聯絡資料傳真至本公司

即期支票
請將支票與您的聯絡資料以掛號方式郵寄至本公司
地址：台北市100中正區博愛路53號7樓

備註
我們提供您快速便捷的送書服務，以及團體購書的優惠折扣。
如單次訂購未達NT1,500，須酌收書籍貨運費用90元，台東及離島等偏遠地區運費另計。
聯絡電話：(02)2383-6000
e-mail: tw_edu_service@mheducation.com

請沿虛線剪下